THE EMERGENCE OF A TRADITION: TECHNICAL WRITING IN THE ENGLISH RENAISSANCE, 1475-1640

Elizabeth Tebeaux

Routledge
Taylor & Francis Group

LONDON AND NEW YORK

First published 1997 by Baywood Publishing Company, Inc.

2 Park Square, Milton Park, Abingdon, Oxon OX14 4RN
711 Third Avenue, New York, NY 10017, USA

Routledge is an imprint of the Taylor & Francis Group, an informa business

First issued in paperback 2017

Library of Congress Catalog Card Number: 96-49606

Library of Congress Cataloging-in-Publication Data

Tebeaux, Elizabeth.
    The emergence of a tradition : technical writing in the English
Renaissance, 1475-1640 / Tebeaux, Elizabeth.
        p.   cm.
    Includes bibliographical references.
    ISBN 0-89503-175-2
    1. Technical correspondence- -History.
T11.3.T43 1996
808'.0666- -dc21   96-49606
                                                                CIP

ISBN 13: 978-0-89503-175-4 (hbk)
ISBN 13: 978-0-415-44232-9 (pbk)

Preface

I have long believed that scholars of technical communication should know the history of their subject. Such knowledge not only provides a sense of the genre's development but also sheds light on current practices. However, until recently, little work had been done to provide a picture of the development of technical communication; in fact, few resources existed that offered an historical perspective of any period in this field. This state of affairs has recently improved, with the publication of books such as Charles Bazerman's *Shaping Written Knowledge: The Genre and Activity of the Experimental Article in Science,* Katherine Adams' *A History of Professional Writing Instruction in American Colleges: Years of Acceptance, Growth, and Doubt,* David R. Russell's *Writing in the Academic Disciplines, 1870-1990: A Curricular History,* and Teresa Kynell's *Writing in the Milieu of Utility: The Move to Technical Communication in American Engineering Programs, 1850-1950,* all of which have filled in gaps in our historical understanding of some aspects of the development of technical and scientific communication. Now we have another important book that helps us understand our history, Elizabeth Tebeaux's *The Emergence of a Tradition: Technical Writing in the English Renaissance, 1475-1640.*

Tebeaux takes as her project the rise and development of printed how-to books during the period from Caxton's establishment of the printing press to the English Revolution. These books cover an astounding number of topics that interested mostly middle-class readers of the period. Newly literate, this class, consisting of merchants, tradesmen, and their wives, as well as other groups, enjoyed increasing wealth and demanded practical education in areas of professional and personal interest. At the same time, knowledge proliferated with the expansion of trade, exploration, and commerce, and, with the growing complexity of knowledge, oral communication became inadequate for preserving evolving bodies of information

in various fields. With the risk of printing, books could be produced to meet the needs of the new market of readers who wanted access to practical knowledge, and, as Tebeaux shows, how-to books on myriad subjects began to appear in the late fifteenth century. Authors wrote and printing presses disseminated volumes on personal medical care, herbals, farming, animal husbandry, gardening, household management, cooking, military science, navigation, and surveying, to cite just the major topics. While some of these books met the needs of the expert reader, others took as their audience the nonexpert who needed to know how to perform some practical activity.

Tebeaux's approach to her material makes the book necessary reading for all people interested in technical communication, whether they be scholars, teachers, or practitioners. She examines Renaissance how-to books through the lens of the central issues of contemporary technical writing. She shows how these early writers created effective books by using the same principles that contemporary writers use, principles such as audience analysis and adaption, readability, the plain style, formatting, and visual aids to clarify technical description.

Tebeaux's analysis of the texts' use of these strategies contributes in many ways to our understanding of technical discourse. One of the most intriguing is her discussion of the evolution of modern formatting conventions that helped make texts readable. The earliest how-to books, she argues, present undifferentiated, dense prose that is difficult to read quickly for needed information. Throughout the Renaissance, authors and printers experimented with techniques to make instructional books more readable, especially to a barely literate readership. Authors and printers drew on the principles of Ramistic logic, with its emphasis on definition and partition of discourse into logical oppositions, to improve the organization of how-to texts. The Ramistic method enhanced the books' overall organization and encouraged the spatial presentation of information by means of page design. Thereafter, books came to objectify information, making it more accessible and easier to use through textual strategies such as indexing, tables of content, white space, overviews, headings, and illustrations, all methods that contemporary technical writers use and instructors teach. By including numerous illustrations of pages from the texts discussed, Tebeaux shows us exactly how these strategies evolved.

Another important contribution to our understanding of the history of technical writing is Tebeaux's analysis of Bacon's contributions to the rise of the plain style. The conventional position holds that Bacon first articulated the principles of the plain style in the seventeenth century and that the members of the Royal Society implemented these principles in scientific discourse. From there it influenced later prose writers. Tebeaux, however, proves that how-to books had used the plain style for a century before Bacon argued for its necessity. Bacon's contribution, therefore, is that he articulated a theory that explained what had long been common practice among technical writers.

Another contribution of this book is that it asks us to reconceive Renaissance genre theory. When scholars discuss the rise of humanism, they usually concentrate on the development of literature, theology, and philosophy. Tebeaux reminds us that humanism also gave rise to technical discourse, with its assumption that humans, if they read the right manual, could successfully accomplish any practical act, whether that be birthing a baby, keeping a set of accounts, surveying one's property, or raising a flock of chickens. Tebeaux also proves that the four traditional genres of Renaissance prose—comedy, satire, epigram, and epistle—inadequately explain the nature of that period's nonfiction discourse. We must add the technical manual to the classification system for a full understanding.

This is an important book, a classic in the field. While its primary audience will be specialists in technical communication, specialists in other areas ranging from literacy theory, history of publishing, theory of textuality, to Renaissance prose will find it instructive. Furthermore, *The Emergence of a Tradition* both fills a scholarly void and provides a model for this kind of historical scholarship. As Tebeaux suggests in her conclusion, researchers can apply her methods to studies of other historical periods, cultures, and types of technical communication.

Michael G. Moran
University of Georgia

Table of Contents

CHAPTER 1

In Search of Our Past

In 1985, Michael Moran wrote that "the history of technical writing has not been written" [1, p. 25]. Adhering to good technical writing practice—to state my primary objective plainly—this book attempts to begin to fill that void. This study will show that technical writing existed in the English Renaissance, that it matured significantly throughout the period primarily as a result of the expansion of knowledge and the rise of print technology, and that the characteristics of the first published English technical books foreshadow characteristics and issues intrinsic to modern technical writing:

- The importance of designing books with readers' comprehension levels in mind.
- The importance of designing books and pages that would be easy to access based on the context in which the reader would use the text.
- The emergence of a structure and a style that would enhance readability and usability of these technical books.
- The development and incorporation of visual aids and the shift from oral to verbal to verbal/visual presentation in the development of technical description as we know it today.
- The triumph of textuality over orality. Throughout the Renaissance, technical writing increasingly textualized oral knowledge. These texts became a means by which knowledge could be generated with greater precision and breadth than oral dissemination allowed.
- The growth of knowledge. Changes in technical writing on a variety of subjects—like medicine, agriculture, and cooking—allow us to see the development of knowledge and the development of the discourse that captured it.

1

Tracking in broad outline the emergence of these basic concepts so inextricably linked to modern technical writing teaching, research, and practice helps illuminate the history of these concerns and the practical concerns that first spurred their emergence. In short, much of what we today call technical communication theory had practical rationale in the sixteenth century.

In attempting to present technical writing as it emerged in the first century of printing in England, I have described much of it with excerpts from the works themselves. My goal in using many examples is to enable readers to see what early English technical books looked like and that these works did indeed lay the foundation for technical writing as we know it today.

My long-range purpose is to show that technical writing not only has a history but also a rich and honorable one shared by the great English Renaissance literary, religious, and philosophical works. For that reason, technical writing, like literature, history, and philosophy, is worthy of study in its own right. Like modern technical writing, Renaissance technical writing differs from other forms of writing not in its cultural or its intellectual origins but in its purpose, the aims of its discourse. It was this purpose, as it shaped content, that gave English Renaissance technical writing its character and made it as valuable to English Renaissance readers as literary, devotional, leisure, and historical reading published throughout the period.

Understanding technical writing of this particular period and any other period provides readers who are teachers and scholars of language a broader understanding of the characteristics of a period than literary or historical studies alone will afford. The English Renaissance was not simply a world of courtiers, drama, political intrigue, political and theological polemic, military and geographical conquests, love poems, catechisms, sermons, and worship aids. It was a world of action and change in medicine, in home and estate management, in agriculture, in business. Understanding the characteristics of technical writing as it existed and changed during the English Renaissance shows us more clearly how people outside the literati and political circles, the majority of English people, lived.

Studies of English Renaissance literature have long focused on great authors, their works, and descriptions of the splendid imaginative, philosophical, and religious prose and poetry that have come to characterize this period. General studies of the English Renaissance have focused on the reigns of the great Tudor monarchs; the social and economic emergence of England as a world power; the emergence of English language, literature, art, and architecture; and the ways in which printing changed the character of England during the sixteenth century. These studies, while focusing on major religious, political, historical, and literary documents, have largely ignored the large body of writing which may be called "technical" writing—how-to books or procedure manuals on a variety of topics: farming, gardening, animal husbandry, surveying, navigation, military science, accounting, recreation, estate management, household management, cooking,

medicine, beekeeping, and silkworm production, to name a few of the major topics covered in how-to books.

If as David Dobrin has suggested, "technical writing adapts technology to the user" [2, p. 247], then these works help define many of the "technologies" used in England during the Renaissance and explain how they were to be implemented in daily living. F. S. Ferguson, in 1913, in recognizing these books, recognized that their aim was "not to speculate, or discuss, or describe, but to give directions how to do something, how to produce something tangible, a practical result for human use or convenience" [3, p. 145]. In short, technical writing was a specific form of instructional discourse that existed in abundance in the English Renaissance and grew in popularity throughout the period.

To teachers, researchers, and practitioners of modern forms of technical writing, my broader goal is to provide a segment of the historical foundation for several basic issues that characterize modern technical writing. Rather than being frequently apologetic for our dealing with a kind of writing that bears the taint of the marketplace and the non-academic world, we can begin to see that technical writing is as much a philosophical product of the Renaissance as is Dante's *Divine Comedy* or Giotto's "The Last Judgment." As the works discussed throughout this study will show, technical writing is a basic form of humanistic expression. It stands as testimony to the Renaissance belief in the power of literacy to transform human existence. With its purpose of helping the individual to perform specific tasks in daily life and work, to live well, to prosper physically and financially, and to acquire knowledge by reading, it asserts the worth and the power of the individual to control human destiny.

Because science in the English Renaissance, particularly astronomy, was only beginning to emerge, I have sought to focus on works that are how-to books about practical topics rather than treatises on causation and the nature of the universe. Because what we might call "science" in the sixteenth century was not even remotely a group of developed fields, books that would today be called science books—books about plant identification, books on anatomy and medical diagnosis, books about pharmaceutical preparation—were how-to books in the English Renaissance. These books, written to help individuals correctly perform processes, anticipate modern scientific works with their emphasis on method and procedure. Many Renaissance technical books can be seen as the first printed English science books. As such, they form the basis of another study, which I challenge someone to pursue—the emergence of English Renaissance scientific discourse. Technical writing in the English Renaissance was, as it is today, writing for the world of work, except that daily life and work in the Renaissance were of one piece.

Thus we can see, through technical books, what tasks were important in English life, how this work was performed, what constituted "work," and why specific tasks were valued. Elizabeth Einstein, in *The Printing Press as an Agent of Change*, noted that the effect of printing has not been measured:

> It is one thing to describe how methods of book production changed after the
> mid-fifteenth century or to estimate rates of increased output. . . . It is quite
> another thing to describe how access to a greater abundance or variety of
> written records affected ways of learning, thinking, and perceiving [4, p. 8].

While the answer to this question—the consequences of printing—is not the focus
of my efforts here, I suggest that the steady, increasing numbers of books published and the increasing sophistication of the content of these books—my focus
here—allow us to infer that advancing knowledge and literacy were changing
people's ways of learning, thinking, and perceiving.

Pollard and Redgrave's *A Short-Title Catalogue of Books Printed in
England . . . 1475-1640*, provides the best, most accessible record of the prevalence and popularity of technical books [5]. While religious works—sermons,
admonitions, polemics, prayerbooks, aids to worship, meditations, homilies, for
example—were clearly the most popular works published in the Renaissance,
many technical books were also popular. As Bennett [6], Wright [7], and Hirsch
[8] have all pointed out, printing as a business was still in its infancy during most
of the Renaissance. Printers could not afford to print what would not sell. As Louis
Wright noted,

> One has only to scan the titles of the *Short-Title Catalogue* to gain some idea
> of the extent and variety of works produced for the literate public in England
> between the introduction of printing and 1640. This huge outpouring of books
> could not have been printed if there had not been an enormous demand from
> the generality of citizens. The publishers of Elizabethan England could no
> more live by the custom of learned and aristocratic readers alone than can
> modern followers of their trade [7, pp. 82-83].

Thus, the substantial numbers of technical books published in England during the
1475-1640 period provide clear testimony to their popularity to readers clearly
dedicated to scripture and religious works.

My decision to focus this study on English Renaissance technical writing stems
from the importance that printing had on textuality and the rapid replacement of
oral transmission with printed transmission of knowledge during the English
Renaissance. Printing in England is inextricably linked to the larger concept of
"Renaissance" as it occurred in England. Thus, the 1475-1640 period, which
begins with the Caxton and ends with the beginning of the English Civil Wars,
offers an ideal crucible for studying the emergence of technical writing. Technical
writing definitely existed before printing, but printing empowered all forms of
written discourse—biblical, philosophical, historical, popular, and technical—in
ways that were impossible when knowledge was confined to manuscript.

All writing, not just technical writing, drew nourishment from the availability of
affordable books and the concomitant rise of literacy. First, expanding literacy
created a demand for books in the vernacular, particularly information and
instructional books that allowed newly literate readers a means of self-education.

Second, population growth and proliferation of knowledge through the printed word meant that knowledge no longer needed to be transmitted solely by oral means or to depend on the oral context to help give meaning to the printed word. Third, in disciplines such as medicine, expanding knowledge became too cumbersome to be passed on orally. Increasing knowledge, empowered by the capabilities offered by improving print technology, transformed the means by which knowledge was communicated and ultimately the readers of this textualized knowledge.

To many earlier studies that have captured glimpses of technical writing's rich and global history as well as the problems inherent in exploring the history of technical writing, I am indebted. Jim Zappen's 1987 analysis of studies focusing on major rhetorical and philosophical issues emphasizes the intellectual depth that a study of the history of scientific and technical discourse yields [9]. Earlier, John Brockmann, in his 1983 article, recognized the major problem in existing studies of technical writing history: "The central problem is that historical research in technical writing has too often been focused only on celebrated authors or scientists as technical writers" [10, p. 155]. Brockmann recognized that examining a "broad spectrum of writers," many of them "uncelebrated," would be "immensely more accurate and meaningful" [10, p. 156]. The approach that focuses on the common man "is a much more accurate gage of historical events because it reveals the day-to-day life of a period or place beneath the flash generalizations and theoretical conceptualizations" [10, pp. 155-156].

Echoing Brockmann, Michael Moran, in his 1985 assessment of published studies in "The History of Technical and Scientific Writing," begins by pointing out that "we have only scattered pieces of scholarship that, when fitted together, do not yet make up a complete picture." Moran urged that his study

> raise questions about why technical and scientific writing changes and develops over time and why individual writers are motivated to write. Finally, it should ask who is the audience for this writing and how audience expectations influence its function and development [1, p. 25].

My method of exploring English Renaissance technical writing has been developed with the perspectives of these three scholars in mind and seeks to address the concerns stated by each. As subsequent chapters will reveal, my indebtedness to the many scholars cited in the review essays is extensive.

Moran's and most recently Rivers's [11] assessments of major scholarship in the history of technical writing have further supported my decision to focus this monograph-length study on the English Renaissance. As Rivers's and Moran's assessments revealed, most of the work in English Renaissance technical writing has been focused on the rise of epistolography. Medieval studies have focused on Chaucer as one of the first technical writers, and Anglo-Saxon technical writing has received scant attention due mostly to lack of accessible sources for study.

Many scholars have found the lure of Bacon's style and method irresistible as his reputation as thinker and writer has continued to dominate seventeenth-century intellectual history. The effects of the Royal Society on style, as seen in studies of Joseph Glanville's style, the rise of the technical paper, the work of Linnaeus and Lavoisier in the development of nomenclature, all show how science, as it emerged decisively in the seventeenth century, has been an easy target for initial explorations of technical writing with an emphasis on the rise of science. The plethora of major scientists—such as Priestley, Darwin, Watson, Crick, and Einstein—and a rich supply of American technical writing sources likely account for the abundance of work in eighteenth-, nineteenth-, and twentieth-century scientific and technical writing history. Rivers's study provides an effective guide to what has been done and remains to be done in various centuries and research eras [11].

Oddly enough, the sixteenth century—the rich years that define the English Renaissance and produced works that reveal the power of print technology—has not been widely explored in terms of technical writing. Science was in its infancy, entangled with superstition and theology, and definitely not approaching the magnitude of true scientific writing that would emerge at the end of the seventeenth century. Thus my study attempts to capture technical writing as it existed in the first period of printed English books, most of them written by uncelebrated writers who were clearly aware that they were writing in a genre that was distinct from history, fiction, and philosophical and theological prose. It attempts to establish the cultural milieu that produced this writing, the audiences for the work, the characteristics of these audiences, the changes in technical writing throughout the period, and the characteristics of technical writing as these emerged and developed from the first years of printing, 1475, to the beginning of the English Civil Wars, 1640. I have deliberately written each chapter to work independently from those before and after it. Thus, some material is repeated. The book can be read sequentially, or specific chapters can be selected.

I have also attempted to design this study based on the guidelines Jimmie Killingsworth and I articulated in 1992: I have located technical writing in the English Renaissance, examined it as it reflects the time in which it was written, examined as many kinds of writing as possible, and summarized the heritage the sixteenth-century English Renaissance writers left to those who would follow [12, p. 27]. In attempting to portray technical writing in the first century of printing, I hope that this necessarily broad-brush visual and verbal presentation of English technical writing during the early years of print technology will provide a more extensive illumination of technical discourse as it foreshadows current issues of today's workplace. If I succeed, this study will provide encouragement for others whose background equips them to reveal the history of technical writing in other eras and cultures.

REFERENCES

1. M. Moran, The History of Technical and Scientific Writing, in *Research in Technical Communication: A Bibliographic Sourcebook*, M. G. Moran and D. Journet (eds.), Greenwood Press, Westport, Connecticut, and London, pp. 25-38, 1985.
2. D. Dobrin, What's Technical About Technical Writing? in *New Essays in Technical and Scientific Communication: Research, Theory, and Practice*, P. Anderson, R. J. Brockmann, and C. Miller (eds.), Baywood Publishing, Amityville, New York, pp. 227-250, 1983.
3. F. S. Ferguson, Books of Secrets, *Transactions of the Bibliographical Society, 12*, pp. 145-176, 1911-1913.
4. E. L. Einstein, *The Printing Press as an Agent of Change: Communications and Cultural Transformations in Early-Modern Europe*, Cambridge University Press, Cambridge, England, 1979.
5. A. W. Pollard and G. R. Redgrave, *A Short-Title Catalogue of Books Printed in England, Scotland, & Ireland, and of English Books Printed Abroad, 1475-1640* (2nd Edition), W. A. Johnson, F. S. Ferguson, and K. F. Panzer (comps.), Two Volumes, Bibliographical Society, London, 1976 and 1986.
6. H. S. Bennett, *English Books & Readers 1475-1557*, Cambridge University Press, London, 1952.
7. L. B. Wright, *Middle-Class Culture in Elizabethan England*, Cornell University Press, Ithaca, New York, 1935.
8. R. Hirsch, *Selling and Reading, 1459-1550*, Harrasowitz, Wiesbaden, Germany, 1975.
9. J. P. Zappen, Historical Studies in the Rhetoric of Science and Technology, *The Technical Writing Teacher, 14*, pp. 285-298, 1987.
10. R. J. Brockmann, Bibliographical of Articles on the History of Technical Writing, *Journal of Technical Writing and Communication, 13*:2, pp. 155-166, 1983.
11. W. R. Rivers, Studies in the History of Business and Technical Writing: A Bibliographic Essay, *Journal of Business and Technical Communication, 8*:1, pp. 6-57, 1994.
12. E. Tebeaux and M. J. Killingsworth, Expanding and Redirecting Historical Research in Technical Writing: In Search of Our Past, *Technical Communication Quarterly, 1*:2, pp. 5-32, 1992.

CHAPTER 2

The Rise of Technical Writing in the English Renaissance

Understanding the emergence and the role of technical writing in the English Renaissance first requires reflection on major events that shaped the Renaissance in England. Traditional characteristics used to define the English Renaissance have stressed the confluence of changes that reshaped English culture from 1475 to 1640, from the advent of printing to the beginning of the English Civil Wars: these included the growth of wealth, the growth of knowledge, the advent of printing by Caxton and his followers. They also led to the emergence of technical writing as a distinct kind of writing in the Renaissance.

EVENTS SHAPING ENGLISH RENAISSANCE CULTURE

The Growth of Wealth

The growth of wealth was crucial to what would later be called "The Renaissance." When feudalism became unprofitable, industry and commerce moved outside estate borders. National interests based on the economic interests of groups of feudal states emerged, and a money economy replaced the local barter system as goods were bought and sold between nations rather than between neighboring feudal states. As Will Durant succinctly evaluated this economic shift, when the serf produced for his lord, he had scant motivation for hard work. But when the free peasant and the merchant could sell their products in the open market, the incentive for personal gain quickened the economic pulse of the nation; the villages sent more food to the towns, the towns produced more goods to pay for purchases, and the exchange of surpluses overflowed the old municipal

limits often controlled by guild restrictions, to cover England and reach out beyond the sea. Some guilds formed merchant companies licensed by the King to sell English products abroad. The British now built their own ships, which they sent to the North sea, the Atlantic coast, and the Mediterranean. Thus, by 1500 the merchant adventurers of England ruled the trade of the North sea and were expanding toward the south [1, p. 109]. As a result of expanding commerce with the Mediterranean, English traders and seamen needed more advanced accounting and banking methods to chart transactions, and thus books explaining these methods.

The expansion of trade also nurtured the rise of the commercial class in England by providing the gold upon which it was based. Moreover, it encouraged within the commercial class a new middle class which valued thrift, ambition, and self-improvement. While the courtier cultivated the advice given by Castiglione and Peacham on becoming a courtly gentleman, the tradesman was pursuing a practical education designed to multiply his accomplishments. This working citizen, who had less time to pursue leisurely learning, advanced his knowledge through reading.

The answer to the demands of the middle-class reader for practical information appeared in the form of the handbook, the manual or guidebook, the Tudor and Stuart version of the self-help instruction book. With the grammar schools and the universities emphasizing classical education to provide literacy, printers soon saw that a market existed for instructional books that covered a variety of practical subjects. In addition to books on personal conduct and guides to letter writing, books also appeared that helped the newly wealthy merchant manage his new country estate [2, pp. 121-122].

The Growth of Knowledge and Education

With the population changing economically and geographically, the need for communicating knowledge by other methods than oral transmission spurred the demand for books and emphasized the value of literacy. At the same time, print technology reduced the costs of books and thus placed textualized knowledge in reach of an increasing audience of middle-class readers. Thus, the rise of commerce, exploration, and trade placed a new premium on education. During the Renaissance, London became famous for its schools, and urbanization helped to fuel the intellectual awakening of the newly literate readers who viewed education as the means of upward mobility. Education, no longer limited to formal institutions such as the grammar schools and the universities, became available in a wide variety of practical short courses.

Edmund Howes's "The Third Vniversity of England" recounts lectures at Gresham College in such practical subjects as arithmetic, swimming, military science, stenography, painting, geography, and navigation [3]. Sir Thomas Gresham, a successful businessman, was one of several English businessmen who

became so impatient with the entrenched scholasticism of the universities that he founded a college on the premise that education should be practical as well as classical. It is not surprising that this emphasis on the practical inspired many works that may be called technical writing. As Wright observed, Gresham's goal was to combine humanistic and utilitarian learning on a level that would be useful to the intelligent citizen: "A precursor of modern university extension courses, Gresham College was the first of the great institutions devoted to popularizing learning for the benefit of the middle classes" [2, p. 65].

The sixteenth century also marked the birth of the importance of pragmatic textbooks in England. Many of these technical books in medicine, navigation, shorthand, and swimming were clearly written for instructional purposes and could be used either by the individual or by the student in a classroom. Desire for learning, attained by private reading or by formal education, persisted in middle-class consciousness throughout the sixteenth and seventeenth centuries. English people firmly believed that education was an instrument of religious, economic, and social salvation. The rediscovery of ancient learning and the opening of new fields of knowledge expanded the interests of all classes. In Elizabethan society, which was still topsy-turvy as a result of commercial expansion and the break-down of feudal institutions, upward mobility became possible [2, p. 78]. Knowledge, available through an increasing number of how-to books, allowed literate English people access to new as well as ancient information.

The importance of learning to the growing numbers of middle-class readers, coupled with the increasing numbers of books printed between 1550 and 1640, suggests that literacy, defined as the ability to read printed works, was increasing. As Bennett argued, printers soon learned to cater to the tastes of the public [4]. The range of books printed showed that the most popular books dealt with religious, homiletic, practical, and instructional topics. Printers provided these in small volumes that would be inexpensive and easy to handle.

The Growth of Trade

The growth of international trade led to the need for books on travel, geography, navigation, interest calculation, and marketing techniques for English business-men attempting to capture foreign markets. The exploratory voyages of the Spanish, French, and English during the 1420-1560 period nearly quadrupled the known surface of the globe. To the English businessman, the value of traditional academic learning declined. Explorers returning from voyages reported the world in ways that differed significantly from Aristotelian accounts. As a result, tech-nical books emphasizing navigation methods, not books on Greek philosophy, became the important means for a successful commercial class to grapple with the new world [1, p. 863]. Silent reading, rather than formal education, became an essential tool to enhance self-education, to enable people to learn about the

findings of explorers, and to supplement the oral transmission associated with on-the-job training.

Science also benefited from the expansion of knowledge. Fostered by exploration and the effects of the Mediterranean Renaissance in medicine, botany, anatomy, mathematics, and accounting, many English readers became interested in the new science. Technical books in medicine, botany, and anatomy reveal the struggles of these disciplines to free themselves from astrology, pure quackery, and blind reverence for Greek tradition.

The Rise of Humanism

The effects of a new economics, trade, knowledge, and printing were inextricably entwined with the rise of humanistic philosophy and technical writing. Humanism, like art, perhaps received its most essential impetus, its justifying philosophy, from the triumph of Aristotelian thought by St. Thomas Aquinas, the late thirteenth-century philosopher who emphasized the importance of the day-to-day world and the power of the human intellect to reveal divine knowledge and to create worldly knowledge for the betterment of humanity. Aquinas's thought is tangential to the initial movement to resurrect the arts, the emergence of the gothic in the arts, and the renewed importance of the transmission of knowledge.

Aquinas, in contrast to Augustine, believed that the fall did not totally debilitate human reason. In Aquinas's view, only the will was corrupted; the intellect was not affected. Men could confidently rely on their own wisdom. The difference between Augustine and Aquinas, as seen in the grace versus nature controversy which pervaded theology until the closing years of the seventeenth century, is aptly exemplified in Raphael's "The School of Athens." In this fresco, Raphael painted Plato with one finger pointing upward, indicating Plato's reverence for the world of Ideals. In contrast, Raphael presented Aristotle with his fingers spread wide and thrust down toward earth, indicating the Aristotelian interest in the real world.

Thomistic thought emphasized the world as real, not just as symbolic of a higher order. Art, which would infuse English technical writing from 1550 onward, became less symbolic and more realistic. Paintings captured the visible world and people in multi-dimensional representation that emphasized both as they actually existed. Byzantine frescoes had portrayed the world as composed of depthless, one-dimensional living beings which were symbols of divine power. In contrast, the Renaissance emphasis on nature as organic (living and changing) infused art with human and natural subjects that had life, movement, and emotion.

Writing began to show this same richness in dimension and detail. Technical books increased in length and complexity of content and clearly evidenced confidence in the power of human intellect to exert more and more control over the world. Comparing books published about 1550 with books published about 1615

strikingly reveals the development of this richness. But, in absorbing the Italian and European Renaissance, English technical books, like theological books, also combined elements of Christianity and the Classics. Many technical instruction books used the preface to sanction the book's contents by allusions to the Church fathers and appeals to Biblical authority. For example, most books about gardening, wound treatment, and animal husbandry incorporated allusions to the teachings of Aristotle, Galen, and Paracelsus. These early forms of technical writing made no effort to achieve objectivity by extricating themselves from classical philosophy and folklore.

Bacon's contribution to scientific method was, like these Renaissance technical manuals, another manifestation of the humanistic belief in knowledge. The Renaissance belief in the power of education, particularly self-education through the printed text, gave people the power to control their lives as far as their individual intellect could carry them. Because technical writing records instructions, descriptions, and processes of people doing things, it is, from one perspective, perhaps the most humanistic of Renaissance humanist writing. As many works will reveal, the study of man as he is and as he can perform work is a characteristic not only of humanist writing but of technical writing as well. Textualizing instructions operates on the belief that the individual, able to read the text, can perform any task.

As Ong has already emphasized in his studies of the rise of textuality [5, 6], the textualizing of information on discrete pages also came to suggest that text was synonymous with truth. Literacy came to mean the power to know truth by accessing the word as it was contained on the printed page. Literacy can thus be seen as the textual equivalent to Michelangelo's statues of men tearing themselves out of the rock. Literacy and text gave people the ability to begin to free themselves from the unknown world of unseen forces operating in nature.

While early Renaissance technical writing recognized the role of God's grace in all work and asked God's blessing on all forms of human endeavor—surgery, accounting, farming, animal husbandry, and navigation, for example—by the closing years of the English Renaissance, more of the technical books focused on the processes themselves and the specific details necessary to perform the processes. Allusions to God declined. By the second decade of the seventeenth century, the expanding content of technical books suggested that with appropriate and sufficient knowledge, the individual could do anything—raise silkworms, bring a child into the world, repair highways, recognize and treat disease. Knowledge acquired through schools and personal reading made the individual's horizons limitless. While oral transmission of information was limited by the individual's memory, textualized knowledge—properly written and displayed—could be read, re-read, and then used for reference. In the true spirit of Aquinas, the individual had become, by way of the text, the measure of things. By command of text man had become, as it were, master of his own destiny.

The Power of Type

The importance of printing to the social and philosophical ferment which would later be labeled Renaissance cannot be overstated or separated from other influences on this fertile period. Printing definitely aided the spread of literacy. Sheer numbers of books, many cheaply printed in small, easy to carry octavos, indicate the existence and growth of a literate reading public eager for works in the English vernacular. The popularity of technical books as well as religious works also suggests that technical writing served a major need in the lives of Renaissance English people.

In short, technical writing as we recognize it today was appearing in substantial quantities in the Renaissance [4]. Our contemporary concern with audience, format and page design, style, organization, and visual aids—as these are chosen to match literacy level and user context—can also be seen as influencing the development of Renaissance technical books. Understanding these early forms of technical writing in English provides modern technical communicators with a sense of our past and the knowledge that current topics and issues that characterize research, teaching, and practice of modern technical writing were also concerns of English Renaissance technical writers.

CHARACTERISTIC TECHNICAL BOOKS OF THE ENGLISH RENAISSANCE

The degree to which fifteenth- and sixteenth-century texts foreshadow the issues and characteristics of modern technical books is uncanny, as the following brief overview will suggest.

Medical Books

One of the largest categories of how-to books that emerged during the Renaissance was the self-help medical book. Paul Slack, in his study of Renaissance vernacular medical literature, notes that medical books "comprised only an eighth of titles, but a quarter of all editions and nearly a third of works of 'pocket' size" [7, p. 247]. Books for non-medical readers dominated the list of medical best-sellers between 1485 and 1604 [7, p. 247]. Many were translations of continental European works, such as Roesslin's *The Birth of Mankinde* (10 editions, 1540-1604), the most popular book for midwives published in England during the Renaissance [8]. Physicians were a luxury that most middle-class English people could not afford, even when available. These little books contained descriptions of various conditions and suggested remedies, usually consisting of herbs. Other self-help medical books emphasized ways of staying healthy, symptoms of potential health problems, suggestions for proper diet, and ways to prepare herbal remedies.

The model for most of these self-help medical books was Sir Thomas Elyot's *The Castle of Helth* (16 editions, 1536-1610), a guide to self-diagnosis, dietary advice, and practical remedies for a host of conditions such as gout, nosebleeds, coughs, and hemorrhoids [9]. Figure 1 shows a page from The Fyrste Boke, which lists foods, herbs, and spices needed to maintain or improve the health of various organs. Other popular books were *The Myrour or Glasse of Helth* (17 editions, 1531-1580) [10] and Andrew Borde's *The Breuiary of Helth* (8 editions, 1540-1582) [11], both written for general readers and physicians.

Taken as a group, these books define the health problems and fears encountered by English men and women of the Renaissance. They show the difficulty medicine had in freeing itself from superstition, theology, and pure quackery in treating common ailments such as fevers, kidney stones, nosebleeds, headaches, boils, pimples, canker sores, jaundices, burns, and plague—to name a few conditions included in the most popular books.

By the closing years of the sixteenth century, however, these books also showed the growth of technical knowledge that had taken place. *The Secrets of Alexis of Piemont* [12], for example, enjoyed twenty-one editions between 1558 and 1605. During that time it increased to four volumes and more than 900 quarto pages and covered a variety of topics in addition to treatment of diseases and complaints. Among the subjects one finds instructions for ironworking, stain removal, and dyemaking; recipes for beauty aids, dentifrices, vinegar, paint, turpentine, soap, and sachet, to give only a very few examples—all of which would be useful to the estate owner who had a variety of responsibilities, including the health care of family and servants.

A second type of medicinal manual was the herbal, which described numerous plants, explained their medicinal uses, and described the expected health benefits each would provide. These ranged from random collections of short verbal descriptions of common plants and their uses, published in small, cheaply printed octavos and quartos, to large well-illustrated folios of more than a thousand pages. Figure 2, facing pages from Askham's *A litle herball of the properties of herbes* (1561?), exemplifies the rude approach used in the early herbals [13]. The right-hand page reads, in part, as follows:

Scabiosa.

Thys is called Scabious, it is hote & dry in the third degre, to drye it ther is no profit inn it, for scabs take the juice of it, vinegar & oile, & boyle them together til thei waxe thick, & kepe it, for it is good for scabbes. For the Emeraudes seeth them in water, then sit ouer it and take the sume of it and use it, & thou shall be hole, stampe it and seeth it in wine and the drinke is good to destroye humours in the stomach, & drinke it eueri dai fasting with Euphrase and thou shalte be hole, and thou shalt neuer haue the pestilence, breding within the while thou do use it, for the lyuer stampe it and seth it in wyne and drynke it.

THE FYRSTE

Thynges good for the heed

☞ Cububes.
Galyngale.
Lignum aloes.
Maioram.
Baulme myntes.
Gladen.
Nutmegges.
Muske.
Rosemarye,
Roses.
Pionye.
Hiſſope.
Spyke.
Camompll,
Mellplote.
Fewe.
Frankyncenſe.

Thynges good for the harte

☞ Cyramome.
Saffron.
Corall.
Cloues.
Lignum aloes,
Perles.
Macis.
Baulme myntes.
Mirobolanes.
Muſke,
Nutmigges.
Rolemarye.

The bone of the harte of a redde dere.
Maioram.
Buglolle.
Borage.
Setuall.

Thynges good for the liuer

☞ Wormewode.
withwynde.
Agrymonye.
Saffron.
Cloues.
Endyue.
Lyuerwoite.
Cyhorie.
Plantayne.
Dragons.
Raylons greate.
Saunders.
Feuelle.
Uiolettes.
Roſe water.
Letyſe.

Thynges good for the lunges

☞ Elycampane.
Hyſope.
Scabiole.
Cyhouſe.
Raylons.
Maydenheare.

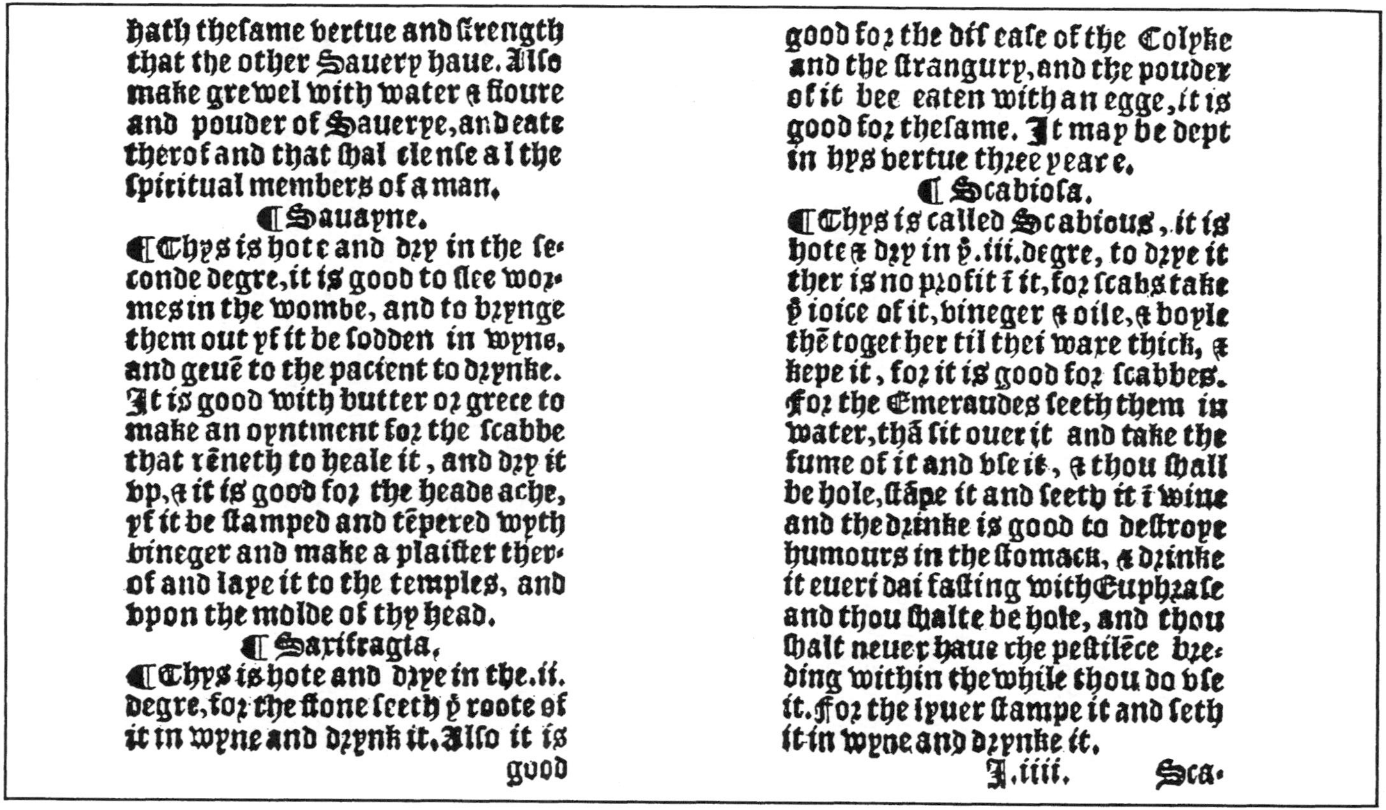

hath the same vertue and strength that the other Sauery haue. Also make grewel with water & floure and pouder of Sauerye, and eate therof and that that clense al the spiritual members of a man.

¶ Sauayne.

¶ Thys is hot and dry in the seconde degre, it is good to slee wormes in the wombe, and to brynge them out yf it be sodden in wyne, and geuē to the patient to drynke. It is good with butter or grece to make an oyntment for the scabbe that tēneth to heale it, and dry it vp, & it is good for the heade ache, yf it be stamped and tepered wyth vineger and make a plaister therof and laye it to the temples, and vpon the molde of thy head.

¶ Saxifragia.

¶ Thys is hote and drye in the.ii. degre, for the stone seeth ý roote of it in wyne and drynk it. Also it is good

good for the dyseease of the Colyke and the stranguty, and the pouder of it bee eaten with an egge, it is good for the same. It may be kept in hys vertue three yeare.

¶ Scabiosa.

¶ Thys is called Scabious. it is hote & dry in ý.iii.degre, to dyve it ther is no profit it, for scabs take ý ioice of it, vineger & oile, & boyle the together til thei ware thick, & kepe it, for it is good for scabbes. For the Emerandes seeth them in water, thā sit ouer it and take the fume of it and bleis, & thou shall be hole, seep it and seeth it i wine and the ryne is good to destroye humours in the stomack, & drinke it euerf daí fasting with Euphrase and thou shalte be hole, and thou shalt neuer haue the pestilece breding within the while thou do use it. For the lyuer stampe it and seeth it in wyne and drynke it.

17

Figure 2. From Anthony Askham, *A lile herball of the properties of herbes* (1561?) [13].

This selection gives a typical example of the information provided by a Renaissance herbal. It also provides a benchmark for comparing early herbals with later ones, such as Gerarde's [14], illustrating the development of knowledge.

The study of herbals is important in the study of the rise of both technical writing and science. In fact, the rise of botanical science can be traced through herbals published in the English Renaissance—from Carey's *Here begynnyth a newe mater* (1525) [15] to Turner's *The First and Second Partes of the Herbal of William Turner* . . . (1568) [16], to Gerarde's *The Herball or Generall Historie of Plants* (1597) [14]. William Turner, probably as true a scientist as the English Renaissance could produce, attempted to describe plants by writing carefully worded descriptions of his visual observations. John Gerarde's *Herball* of over 1400 folio pages included woodcuts, physical descriptions, locations and time for selection, and uses. While neither Turner nor Gerarde even remotely approached a classification system for plants or possessed a nomenclature for description, their attempts at verbal precision and visual accuracy of plant illustrations were clearly forerunners of the birth of botany as a science.

Uroscopy books, a third type of self-help medical book, described how the color, odor, and density of urine might be used to indicate a particular problem. The descriptive terminology, although highly unscientific from today's medical perspective, was a landmark in the advent of technical description. At least four uroscopy books were printed in a total of twenty-one editions during the sixteenth century: *Here Beginneth the Seynge of Urynes* . . . (13 editions, 1525-1562) [17]; *The vrinal of physick* (4 editions, 1547-1599) [18]; *The iudycyall of vryns* (1527) [19]; and *The Judgment of All Urynes* (1540) [20].

In addition, there were a number of specialized books for physicians—procedure manuals describing treatments for disease or surgical methods. These texts were usually printed in large folio or quarto and contained a substantial number of drawings to illustrate parts of the human anatomy, surgical procedures, medical treatment, and pharmaceutical preparations. The most important books, published in the sixteenth century or in the first decades of the seventeenth century, were either written by renowned English physicians or were translated from the highly respected works of French surgeons, such as Guillemeau and Paré, or Italian anatomists, such as Vesalius. An example of the former was Thomas Gale's *Certaine workes of chirurgerie* [21]; of the second, Jacques Guillemeau's *The Frenche Chirurgerye* [22].

Anatomy books, such as Alexander Read's *A Description of the Body of Man* [23] and Nicholas Udall's translation of Thomas Gemini's *Compendiosa Totius Anatomie Delineatio Aere* [24], showed the influence of Vesalius, da Vinci, and other major Italians who were responsible for the birth of modern anatomy and the eventual demise of Galen as the revered authority on human physiology. Both works show the advancing state of technical description of human anatomy available in England by the closing years of the Renaissance. The

multi-dimensional drawing shows the effects of da Vinci's anatomical sketches, which will be discussed in Chapter 6.

Books on Farming and Animal Husbandry

Books on farming and farm animals form another major category of Renaissance technical writing. The importance of agriculture and meat production in Renaissance England is clearly evident in the numbers of technical books on farming, surveying, and caring for cattle, sheep, horses, and fowl. Fitzherbert's *Booke of Husbandrie* [25] enjoyed twenty-one editions between 1523 and 1598, with more information added to each edition. The book consisted of four sections. The first section contained instructions on sowing, tilling, weeding, manuring, and harvesting. The second section emphasized the care of sheep, goats, cattle, horses, swine, and bees. The third dealt with gardening, orchard maintenance, herb and fruit production, distillation of fruits, as well as medicinal preparations made from fruit and herbs. The last section presented rules for household governance—duties of various servants; instructions on how to live happily and virtuously; instructions for wives, whose general duties included care of chickens, geese, ducks, pigeons, pheasants, turtles, partridges, and swans.

Books like Fitzherbert's show what work was deemed important as well as the best processes for performing that work. Because he was an experienced farmer, Fitzherbert provided readers with a sense of the level of knowledge available and advocated in caring for cattle and poultry.

Cheape and Good Hvsbandry . . . [26] by Gervase Markham, first published in 1614, provided even more extensive advice than did Fitzherbert on horses, cattle, sheep, goats, conies, poultry, geese, turkeys, water fowl, hawks, bees, and fish. In contrast to Fitzherbert's *Booke of Husbandrie* with its heavy emphasis on right living, *Cheape and Good Hvsbandry* replaced the traditional didacticism with detailed instructions for treating specific diseases. If we compare these two books, which were published several decades apart, we can see how Markham's book, with its more technical, factual approach to husbandry, has extricated animal care from superstition much more successfully than Fitzherbert was able to do.

Another popular work was Gervase Markham's *Covntrey Contentments* [27], which was divided into two books. The first book, written for men, contained a variety of instructions on caring for horses, recognizing and curing their ailments; instructions for hunting and riding; instructions on caring for hawks and greyhounds; instructions for shooting longbows and crossbows; and instructions for bowling and tennis. The second book, written for the "English Hus-wife," contained instructions for treating a variety of ailments, in addition to listing recipes for salads, meats, pastry, jelly, desserts; instructions for growing and harvesting hemp and flax; instructions for caring for cows, for milking, for making cheese and butter; and instructions for making beer. Both books provide a descriptive view of life on the English Renaissance estate and knowledge needed by new

estate owners to care for land and livestock. Books like Markham's and Fitzherbert's also describe the responsibilities of men and women and the expected roles of each as defined by Renaissance social standards.

Another popular book by Fitzherbert was his book on surveying: *Here begynneth . . . the surueyeng and improumetes* [28]. The work enjoyed twelve editions and reflected the Englishman's desire to protect his land. The book contained useful models for the preparation of documents and much historical information and advice on procedure in manorial courts.

Both Fitzherbert and Markham were among a group of committed technical writers whose work would have qualified them for membership in a Renaissance technical communication society (had one existed). Other writers who would have been considered technical writers include Sir Thomas Elyot, Leonard Mascall, Thomas Blundeville, Thomas Hill, Reginald Scot, and John Partridge. All wrote other kinds of works—literary, historical, educational, and religious—and illustrate superb skill in adapting content to audience; but Fitzherbert, Markham, Mascall, and Hill were committed technical writers.

In contrast to Fitzherbert's and Markham's books on farming and husbandry stands Tusser's *Five Hundred Points of Good Husbandry* [29]. Published twenty-four times between 1557 and 1638, the rhyming poetic stanzas, apparently designed to be read aloud to illiterate readers who could memorize them, contained planting and harvesting advice for every month of the year plus advice to the good "huswife" on brewing, malting, baking. The popularity of Tusser's book, which was much less technically accurate than works by Markham and Fitzherbert, serves as a useful reminder that literacy was not uniform. Although the increasing number of books and the increasing technical complexity of books attest to the general increase of literacy, the existence of information books written in poetic stanzas suggests that the oral tradition was still an important means of transferring information. The number of editions of Tusser's book seems to indicate that a segment of readers still found its combination of folklore and astrology in verse more appealing than either Fitzherbert's or Markham's more technical approach to farm management.

Other books included Leonard Mascall's *The first booke of cattell* (9 editions, 1557-1633) [30], *A profitable boke declaring dyuers approoued remedies to take out spots and staines* (3 editions) [31], and *A Booke of the Arte and maner, how to plante and graffe all sortes of trees* (9 editions, 1569-1579) [32]. His advice on grafting was direct and concise. In Figure 3, instructions for grafting begin beneath the drawing of the tree:

> This figure sheweth, how all Vynes should be proined and cutte, in a convenient time after Christmas, that when ye cut them, ye shall leaue his braunches very thynne, as ye see by this fygure: ye shall neuer leaue aboue two, or three leaders at the heade of any principall braunch ye must also cut them of[f] inn the mydst betweene the knots of the yung [s]cions, for those be

the leaders which will bring the grape, the rest & order ye shall understand as foloweth.

Books on Gardening

The number of books on farming and animal husbandry is rivaled only by the number of books on gardening. These books show the Englishman's commitment to designing mazes, planting and cultivating orchards, beekeeping, raising kitchen

Figure 3. From Leonard Mascall, *A Book of the Arte and Maner, how to plante and graffe all sortes of trees* (1572) [32].

gardens, and using plants for medicinal purposes—all written with the persistent awareness of the moon's generative powers. Gardening books provided detailed instructions for planting, mulching, irrigating, as well as grafting trees.

Thomas Hill was perhaps the most prolific writer on gardening, with three books: *A most brief and pleasaente treatise, teaching how to dress a garden* (8 editions) [33], *The Gardeners Labyrinth* (5 editions) [34], and *The Arte of Gardening* [35] included instructions for maintaining gardens and mazes, planting according to astrological signs, preparing medicine from plants, beekeeping and collecting honey, and forecasting the weather. Reginald Scot's *A Perfite platforme of a Hoppe Garden . . .* (3 editions, 1574-1578), as its title implies, covered instructions for beginning a hop garden, recognizing and treating diseases, cultivating the hop garden, and harvesting, drying, and packing hops [36]. The book provided written instructions along with woodcut illustrations depicting workers tending the hops at various stages. Lawson's *A New Orchard and Garden* (5 editions, 1617-1637) covers planting, tending, and harvesting and describes all garden types, their design, preparation, and care [37].

Gardening books that contained two sections, one written for men and one written for women, such as *A New Orchard and Garden,* show differences that indicate differences in roles as well as in education level. However, like the books on farming that also contain sections for men and for women, these gardening books do not suggest that the writers believed women readers to be any less literate than men [33]. Omission of Greek and Latin allusions in the sections for women does suggest that Lawson knew that only his male readers would likely have attended grammar schools, where Greek and Latin were part of the standard curricula. Otherwise, sentence structure and use of common gardening terminology in the two sections show little if any differences [35].

Books on Household Management and Cooking

Books on household management, written for middle-class women, show many of the technologies developed for use in the English household. Because McDonald's, Pizza Hut, and the corner deli were not available options for the harried housewife responsible for feeding family and servants, cooking and food preservation were important issues. Development of cooking and household management from 1500 to 1640 can be seen in the increasing variety and complexity of recipes for food and home medical remedies. The first extant printed English cookbook, a mere nineteen pages published by Pynson in 1500 [38], contains only basic recipes for meats and breads. By 1615, these books were 60-100 pages long and contained instructions for making sachets, perfumes, dentifrices, and preservatives in addition to elaborate recipes for breads and confections. That knowledge was increasing even in cooking is evident from

changes in measurement: handfuls changed to quarts and cups; "several" became a specific number; the amount of salt or pepper began to be listed more precisely.

These books describing household management again remind us that women were responsible for the medical treatment of family, servants, and friends. Many household management books contained recipes for "Medicines, Salues, for Sundry Diseases" in addition to instructions for preserving peaches, quinces, oranges, lemons and rose leaves; instructions "For the Woormes," "To stench bleeding at the Nose," "To know whether a Child hath the Wormes, or no."

John Partridge's *The widdowes Treasure* (10 editions, 1582-1639) was a conglomeration of recipes for foods and medicines: for example, "For the Ague," "For the Tooth-ache," "To keepe Venison fresh a long time," "To keep it from rotting after it is new slaine" [39]. Another of his books, *The treasurie of commondious conceits, and hidden secrets, and may be called the huswives closet of healthfull provision*, enjoyed twelve editions between 1573 and 1673 [40]. This little book provides extensive discussion of household preparations, medicinal oils and waters, herbal preparations, and medical conditions. The final section includes a substantial discussion of problems associated with pregnancy. Despite the conglomerative nature of the "recipes," the *Short-Title Catalogue* records over three dozen books of this type. The number of editions and longevity of many of these books in addition to the number of books published suggests that a demand existed for these books, and printers were happy to fill it.

Books on Recreation

The English gentleman's love of horses can be seen in the large number of books on raising and caring for horses. In addition, there were books on raising and caring for falcons and hawks, books on hunting, fishing, and trapping, and books on swimming, archery, and fencing.

Thomas Blundeville's *The fower chiefyst offices belonging to horsemanship* (6 editions) [41] and Markham's four books on horsemanship were the most popular. The latter's *A discource of horsemanshippe* by itself enjoyed five editions (1593-1603) [42]. Other books on sports included Mascall's *The Booke of fishing with Hooke & Line* . . . [43] and George Turberville's *The booke of faulconrie or hauking* [44]. Figure 4 shows a page from *The Booke of fishing with Hooke & Line*, which also contains a section on how to make animal traps. Mascall provided a drawing and an operational definition of each trap.

Falconry was the sport of kings in the Middle Ages, but during the Renaissance it became a sport for English country gentlemen. The high-quality copper plates used in many of the beautiful books on hawking, translations from the French, show the influence of the French Renaissance. Markham also wrote a book on how to catch birds in nets, *Hungers prevention: or, the whole arte of fowling* [45], although books on raising and caring for falcons were the most numerous. *The Boke of St. Albans*, a book of hawking by Berners, first published in 1475, includes

72   The Booke of

The fall for Rats or other vermine.

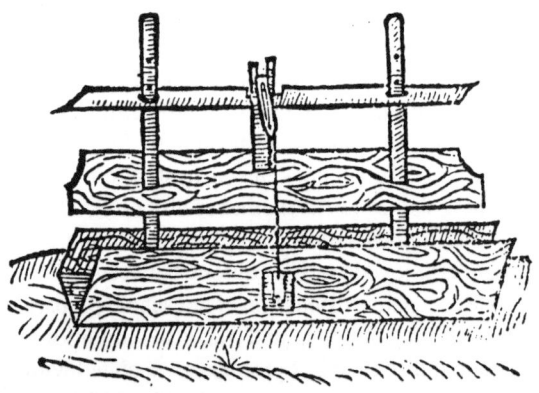

Tᴴis engine is called the Ratte trappe or fall, which is
made with a thicke bottome borde, and two thinner
bordes on both sides, and there is two staues set fast thorowe the bottome borde, then the fall must be a thicke bord
and heauie withall, and at the endes thereof must your
staues goe thorow easely to fall and rise, which two staues
haue holes aboue, which staues must also goe thorowe the
long bridge aboue, and at the holes ye must put in two pins
to holde vp the sayde bridge. Then must ye set fast another
staffe in the middest of the fall, with a latch in the toppe
thereof loose set to fall vp and downe: which latch must haue
a string, which string commeth downe to the bridge beneath, with a small clicket fastened thereunto: and the
bridge is fastened beneath on the backeside or borde, an inch
from the bottome borde, and so it is done. Ye may make
them to take water Rats in setting them in the water, in
the sides of your ponds and riuers, and bayted with carion,
but then ye must set rowes of short nayles vnder the fall
planke, and those will stay either ratte or other fish, if they
goe through it, and put downe the bridge.

The

Figure 4. From Leonard Mascall, *The Booke of fishing
with Hooke & Line* . . . (1590) [43].

sections on raising falcons [46]. By 1614, the work had appeared in twenty-two editions and had changed from a beautifully illustrated folio to a compact quarto.

Beyond horses, falconry, and fishing, books appeared about other sports. There were books of instruction on fencing, such as *Pallas Armata, The Gentlemans Armorie*, "Wherein the right and genuine use of the Rapier and of the Sword, as well against the right handed as against the lefthanded man is displayed" [47]. Christopher Middleton translated a book on swimming, *A Short introduction for to learne to Swimme*, which used text with illustrations to explain the basic swimming strokes and turns in water [48]. Figure 5 shows a page from this book.

Books on Military Science and Navigation

The variety of books on military science and navigation emphasize the importance of conquest to Renaissance Englishmen. There were books on how to build fortifications, many of these used in short courses at Gresham College; books on navigation methods; books describing navigational instruments and the procedures for their use, including the compass, the hemisphere, and the pantometer. Many of these contain some of the finest illustrations available in Renaissance technical books. (See, for example, Figure 18 in Chapter 6, p. 214, which shows a copperplate engraving of the pantometer from a book on the use of navigational instruments.)

John Bingham's *The Art of Embattailing an Army* (1629) provided instructions for preparing soldiers for battle as well as how to conduct a battle [49]. The book is carefully partitioned into specific steps. Instructions with explanatory steps are accompanied by integrated fold-out drawings of military formations. Books on fortification, such as Richard Norwood's *Fortification or Architecture Military* (1639) [50], provided detailed written instructions for building fortifications. Edward Cooke's *The Prospectiue Glasse of Warre, Shewing You a Glimpse of Warre Mystery, In Her Admirable Stratogems* gives illustrated instructions for deploying troops and cannon [51].

Because these instruments had to be made by hand, books were needed to explain how to make the measuring instruments needed in building fortifications. Iohn Blagrave's *A Booke of the Making and Vse of a Staffe, Newly Inuented by the Author* (1590) explained its purpose as: "and at this time publisyhed for the speciall help of shooting in great Orjdinance, and other millitarie seruices, and may as well be imployed by the ingenious, for measuring land" [52]. William Bourne, who wrote several military science instructional books, devoted an entire book to *The Arte of Shooting in Great Ordnaunce* (1578) [53].

Books on navigation provided descriptions of coastlines, sea routes, and navigational instruments. Those that appeared in the early years of the seventeenth century emphasized the use of navigational instruments and included tables of solar declination. A major book in this category was *The safegard of Sailers* (9 editions, 1581-1634), which gave general locations of land masses, drawings of

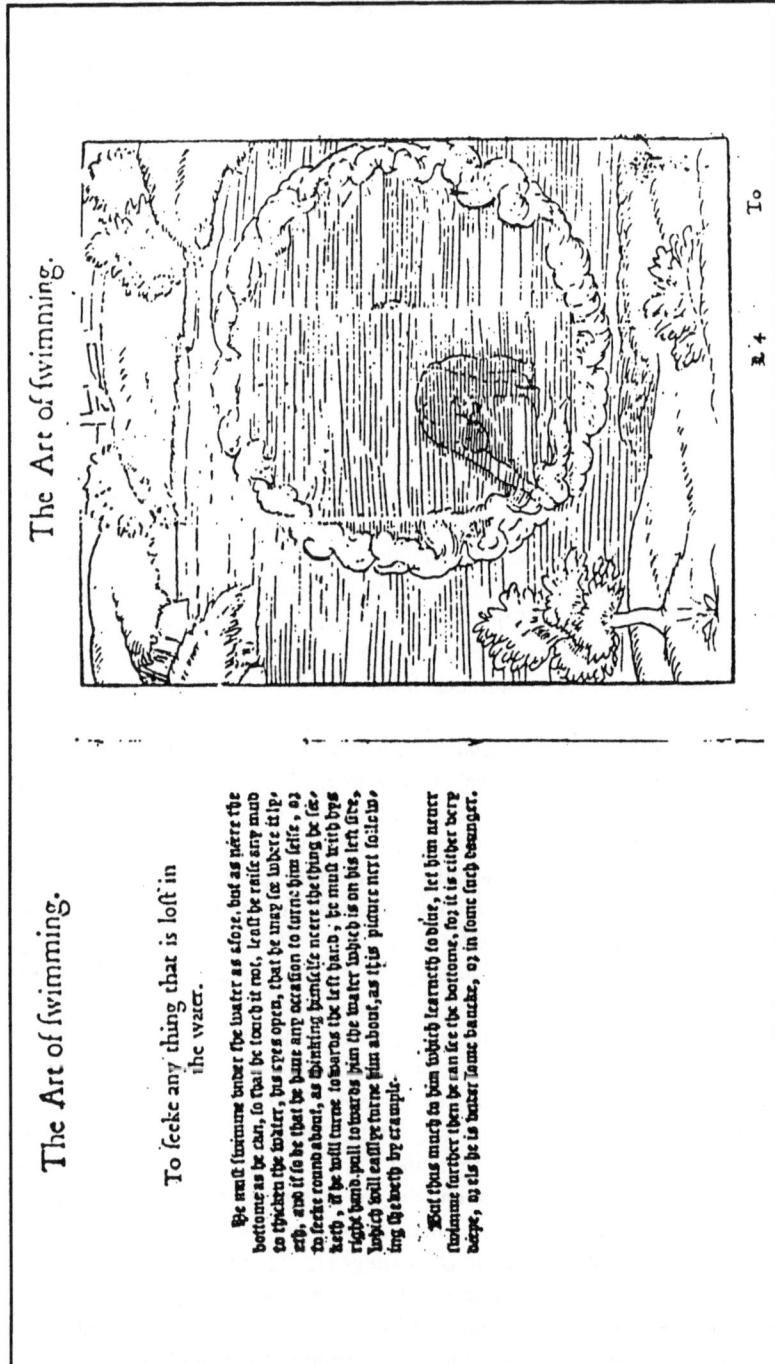

Figure 5. From Christopher Middleton (trans.), *A Short introduction for to learne to Swimme* (1595) [48].

coastlines, and instructions for sailing in terms of days and hours [54]. Figure 6 shows a page from this book. By following instructions, sailors could use the drawings of prominent structures near shore to help them identify their location.

One of the earliest of English navigational books, *Mappa Mundi* (1535?), warns "And also hell is in the myddes of Affryke vnder the earth, and is set in the myddes of the erth, as a karnell of an apple is in the middes of the apple, for as moch earth is there aboue as beneath" [55]. An examination of navigation and geography books shows how pervasively the growth of knowledge affected these fields, as it had affected medicine. Indeed, the effect of knowledge, exploration, and instrument navigation on these works is even more startling than the changes in herbals. As we might expect, translations of Spanish navigation books were popular and vastly superior in technical content to native English works.

Other Technical Books

Other books covered descriptions of surgical instruments, instructions for repairing highways and building ponds, books on designing garden mazes and stained glass windows, and books on carpentry. Many are clear forerunners of modern specification writing. Their precision in description increased sharply during the 1560-1630 period, and these books also bear testimony to the growth of knowledge in all fields and the ability of printers to use type in interesting ways to present this knowledge in textual form.

For example, Thomas Proctor's *A Profitable Worke to this Whole Kingdome. Concerning the mending of all High-ways also for Waters and Iron workes* (1610) included exactly what its title suggests—written instructions with drawings that show how highways paved with rocks and bricks should be repaired [56]. Figure 7 shows a page of this book that illustrates one method of arranging rocks. *A Booke of Svndry Dravghtes* (1615) included designs and instructions for making designs for stained glass windows, instructions for annealing in glass, and recipes for making colored glass [57]. *The Merchants Avizo* . . . (7 editions, 1589-1632), a popular instruction book for seagoing merchants, included model business letters and advice on forms, procedures, and business policies to help the English seaman do business in foreign countries [58]. *The Carpenters Rvle to measure ordinarie Timber* was written for craftsmen who had a basic knowledge of arithmetic but who needed to know how to measure wood so as to avoid waste [59].

There were also books on silkworm production, a job handled by women until the end of the sixteenth century. Like books on farming, these books on raising and harvesting silkworms move from instructions captured in ballad stanzas in T.M.'s *The Silkewormes, and their Flies* [60], first published in 1599, to works such as DeSerres's *The Perfect Vse of Silok-Wormes, and their benefit* [61], published eight years later in 1607. The increase in knowledge about silkworm production during this period is clearly evident. Drawings from William

of Sailers. 14

Item, when you come nigh the second Beacon, there is a good roade for a westerly wind. And then haue you Langh-

Langhebooorde

woerd Church in Gotirland, and if you will go farther, then set your course southeast to the Buy or tunne vpon the Wlacke. And when you are within that Buy, then go with the west shore in three or four fatham, and so you shall come now & then hard by groundland, vntill you come to Blexhemkerk. There be in the roade commonlis such ships as are outwards bound, and doo tarrie for a wind.

Item, when Newwarke lies southeast and bysouth of you, then haue you both the capes in one.

Blexhem Church.

Item, when you haue Menser Church south from you, then are you open before the Wleser.

Item, when the steeples of Borkum beare south from you, then you are open thwart of the easter emes.

Item, when the steeples beare due northeast of you, then you are within both the Juster Keuens, in ye wester emes.

Item, when the hedlands of Kottum doo lie southeast and byeast from you, then they are in one.

Item, when the hedlands lie south southeast from you, then they are both in one.

¶ *The going into the Iade.*

Item, if you will go into the Iade, when you are past the red sand, beare in southwards, euen as far as you may,

C. g.   and

Figure 6. From Robert Norman, *The safegard of Sailers* (1584) [54].

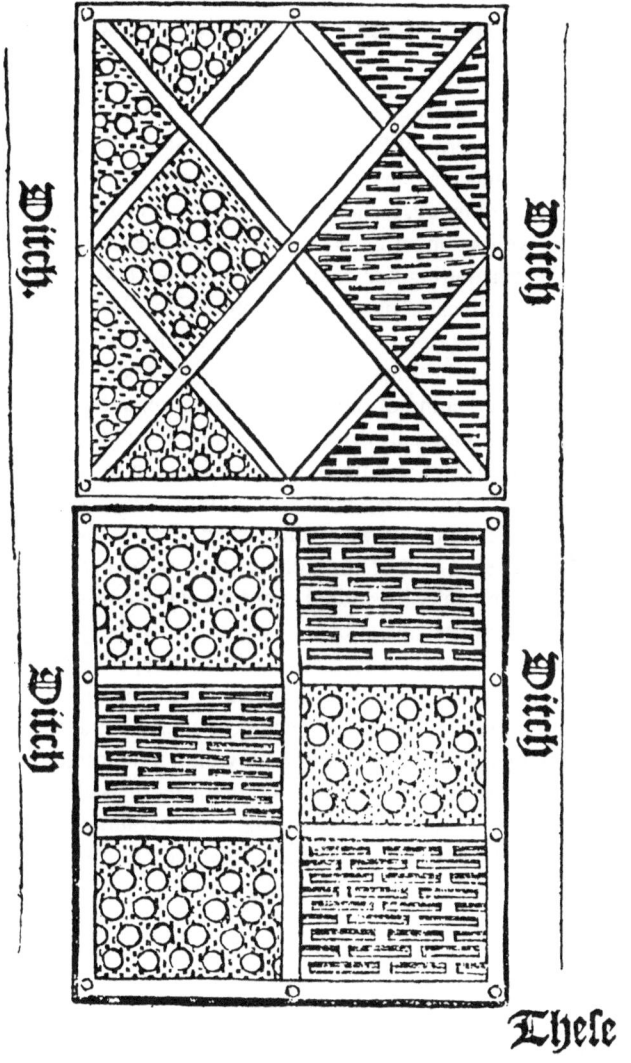

Figure 7. From Thomas Proctor, *A Profitable Worke . . . Concerning the mending of all High-ways . . .* (1610) [56].

Stallenge's *Instrvctions for the increasing of Mulberie Trees* . . . [62], published in 1609, enable us to visualize some of the technology used in breeding silkworms, such as special rooms with racks and shelves. The drawings in this work are supported by correlating verbal descriptions of the stages in the breeding process.

THE VALUE OF HISTORICAL TECHNICAL BOOKS

A study of these technical books, particularly of the range of their content, provides any student of language a broad array of benefits. First, these books suggest the major concerns of English people during the Renaissance—in health, home, commerce, and conquest. Second, read against a general knowledge of English Renaissance history, technical books provide insights into the literacy of middle-class as well as upper-class readers. Third, they track the demise of the oral tradition in transmitting knowledge when printed text could embody extensive, detailed, and increasing quantities of information. Fourth, the emergence of technical books from the first years of English printing to the outbreak of the Civil Wars in 1640 reveals the importance of technical books to English men and women. Fifth, the beginning of objective presentation and style—meaning writing, particularly instructional writing, that was attempting to extricate itself from folklore, superstition, and theology—can be seen in many of these works. As content based on observed behavior or processes increased, the role of metaphysics and pseudometaphysics diminished. Finally, these books reveal the emergence and development of technology in work-related tasks as well as in fields that would eventually be studied under the science rubric.

The extensiveness of technical books, suggested by the general description in this chapter, allows us to include these books in a separate genre that provides another avenue for the study of the development of modern English as it was used for instrumental purposes. These early technical books thus allow modern readers to study the emergence of a genre, while providing new insight into the study of technical discourse as it reflects the intrinsic character of a culture. As we examine some of these books in more depth, particularly the evolution of technical writing as it emerged during the sixteenth and early seventeenth centuries, we will focus on many of these works from five different perspectives, which form the topics of the next five chapters.

Chapter 3 illustrates both writers' and printers' distinct awareness of organization and page design as these facilitated the readability of texts designed to be used for rapid reading and ready reference. Differences between page design in technical books and in philosophical, literary, and religious books suggest that technical writers also knew that technical books would be read rapidly and for information and reference, while other books would be read slowly and leisurely.

Chapter 4 shows how technical writers were aware of the interests, content needs, and reading comprehension levels of the readers they wished to attract.

Technical writers knew that different audiences existed for technical books than for many literary and philosophical works.

Chapter 5 argues that the development of plain style was concomitant with the rise of pragmatic English and that technical writers, by the early years of the seventeenth century, saw the value of a style that was precise, direct, and succinct.

In Chapter 6 I discuss the evolution of illustration and visual aids in the emergence of technical description that by the seventeenth century had obscured the effects of orality in practical discourse.

Finally, in Chapter 7 I discuss the legacy of the Early Renaissance technical writers and suggest further research to fully explore these first printed English technical books.

REFERENCES

1. W. Durant, *The Reformation, The Story of Civilization, Part IV*, Simon and Schuster, New York, 1957.
2. L. B. Wright, *Middle-Class Culture in Elizabethan England*, Cornell University Press, Ithaca, New York, 1935.
3. E. Howes, The Third Vniversity of England, Or a Treatise of the Fovndations of All the Colledges, Ancient Schooles of Priviledge and of Hovses of Learning, and Liberall Arts, Within and About the Most Famous Cittie of London, in *The Annales or General Chronicle of England*, I. Stow, London, 1615 [STC 23338].
4. H. S. Bennett, *English Books & Readers 1475-1557*, Cambridge University Press, London, 1952.
5. W. J. Ong, *Orality and Literacy: The Technologizing of the Word*, Methuen, London, 1982.
6. W. J. Ong, *Ramus, Method, and the Decay of Dialogue*, Harvard University Press, Cambridge, Massachusetts, 1958.
7. P. Slack, Mirrours of Health and Treasures of Poor Men: The Uses of Vernacular Medical Literature of Tudor England, in *Health, Medicine and Mortality in the Sixteenth Century*, Cambridge University Press, Cambridge, England, pp. 237-274, 1979.
8. E. Roesslin, *The Birth of Mankinde, Otherwyse Named the Womans Booke*, T. Raynald (trans.), London, 1598 [STC 21160].
9. T. Elyot, *The Castle of Helth*, London, 1536-1539 [STC 7643].
10. T. Moulton, *The Myrour or Glasse of Helth*, London, 1580 [STC 18225].
11. A. Borde, *The Breuiary of Helth*, London, 1547. Reprinted in *The English Experience*, Number 362, Da Capo, New York, 1971.
12. P. Alessio, *The Secrets of Alexis of Piemont*, W. Ward (trans.), London, 1558, 1595.
13. A. Askham, *A litle herball of the properties of herbes,* London, 1561? Reprinted in *The English Experience*, Number 843, Theatrum Orbis Terrarum, Amsterdam and Norwood, New Jersey, 1977.
14. J. Gerarde, *The Herball or Generall Historie of Plants*, Two Volumes, London, 1597. Reprinted in *The English Experience*, Numbers 660A and 660B, Theatrum Orbis Terrarum, Amsterdam and New York, 1974.

15. W. Carey, *Here begynnyth a newe mater: the which is called an Herbal*, London, 1525.
16. W. Turner, *The First and Second Partes of the Herbal of William Turner, Doctor of Physick Lately Ouersene Corrected and enlarged with the Thirde Parte Later Gathered and Newe Set Oute with the Names of the Herbes in Grece Latin English Duche Frenche and in the Apothecaries and Herbaries Latin with the Properties Degrees and Natyurall Places of the Same*, London, 1568 [STC 24367].
17. *Here Beginneth the Seynge of Urynes of All the Couloures that Urynes Be of with the Medycines Annexed to Every Uryne and Every Uryne His Urynall Much Profitable for Euery Man to Know*, W. Powell, London, 1562 [STC 22153].
18. R. Record, *The vrinal of physick*, London, 1547 [STC 20816].
19. *The iudycyall of vryns*, P. Treveris, London 1527 [STC 14836].
20. D. Smyth, *The Judgment of All Urynes: And for to Know the Manes from the Womannes and Beastes Both from the Manes and Womans with the Coloure of Euerye Dryne*, London, 1540 [STC 24595].
21. T. Gale, *Certaine workes of chirurgerie*, Four Parts, London, 1563 [STC 11529].
22. J. Guillemeau, *The Frenche Chirurgerye*, London, 1597 [STC 12498].
23. A. Read, *A Description of the Body of Man*, London, 1634 [STC 20783].
24. T. Gemini, *Compendiosa Totius Anatomie Delineatio Aere Exarata*, N. Udall (trans.), London, 1553 [STC 11716].
25. J. Fitzherbert, *Booke of Husbandrie*, London, 1598. Reprinted in *The English Experience*, Number 926, Theatrum Orbis Terrarum, Amsterdam, 1979.
26. G. Markham, *Cheape and Good Hvsbandry For the well-Ordering of all Beasts, and Fowles, and for the generall Cure of their Diseases*, London, 1614. Reprinted in *The English Experience*, Number 139, Theatrum Orbis Terrarum, Amsterdam and New York, 1969.
27. G. Markham, *Covntrey Contentments, in Two Bookes*, London, 1615. Reprinted in *The English Experience*, Number 613, Theatrum Orbis Terrarum, Amsterdam and New York, 1973.
28. J. Fitzherbert, *Here begynneth the ryght frutefull mater: and hath to name the boke of surueyeng and improumetes*, London, 1523. Reprinted in *The English Experience*, Number 657, Theatrum Orbis Terrarum, Amsterdam and Norwood, New Jersey, 1974.
29. T. Tusser, *Five Hundred Points of Good Husbandry*, London, 1580. Reprinted, G. Griegson (ed.), Oxford University Press, Oxford, 1984.
30. L. Mascall, *The first booke of cattell*, London, 1587 [STC 17580].
31. L. Mascall, *A profitable boke declaring dyuers approoued remedies to take out spots and staines*, London, 1583.
32. L. Mascall, *A Booke of the Arte and maner, how to plante and graffe all sortes of trees*, London, 1572. Reprinted in *The English Experience*, Number 697, Theatrum Orbis Terrarum, Amsterdam and Norwood, New Jersey, 1974.
33. T. Hill, *A most brief and pleasaente treatise, teaching how to dress a garden*, London, 1563 [STC 13490].
34. T. Hill, *The Gardeners Labyrinth*, London, 1577 [STC 13485].
35. T. Hill, *The Arte of Gardening*, London, 1608. Reprinted in *The English Experience*, Number 936, Theatrum Orbis Terrarum, Amsterdam and Norwood, New Jersey, 1979.

36. R. Scot, *A Perfite platforme of a Hoppe Garden, and necessarie Instructions for the making and mayntenaunce thereof*, London, 1574. Reprinted in *The English Experience*, Number 620, Theatrum Orbis Terrarum, Amsterdam and New York, 1973.

37. W. Lawson, *A New Orchard and Garden*. London, 1623 [STC 15329].

38. [Book of Cookery], R. Pynson, London, 1500 [STC 3297].

39. J. Partridge, *The widdowes Treasure*, London, 1595 [STC 19434].

40. J. Partridge, *The treasurie of commondious conceits, and hidden secrets, and may be called the huswives closet of healthfull provision*, London, 1573 [STC 14425].

41. T. Blundeville, *The fower chiefyst offices belonging to horsemanshipps*, Four Parts, London, 1565 [STC 3152].

42. G. Markham, *A discource of horsmanshippe*, London, 1593 [STC 17346].

43. L. Mascall, *The Booke of fishing with Hooke & Line, and of all other instruments thereunto belonging. Another of sundrie Engines and Trappes*, London, 1590. Reprinted in *The English Experience*, Number 542, Theatrum Orbis Terrarum, Amsterdam and New York, 1973.

44. G. Turberville, *The booke of faulconrie or hauking*, London, 1575 [STC 24324].

45. G. Markham, *Hungers prevention: or, the whole arte of fowling*, London, 1621 [STC 17362].

46. J. Berners, *The Boke of St. Albans*, St. Albans, 1486. Reprinted in *The English Experience*, Number 151, Theatrum Orbis Terrarum, Amsterdam and New York, 1969.

47. *Pallas Armata, The Gentleman Armorie*, London, 1639 [STC 3].

48. C. Middleton (trans.), *A Short introduction for to learne to Swimme*, London, 1595 [STC 6840].

49. J. Bingham, *The Art of Embattailing an Army*, London, 1629. Reprinted in *The English Experience*, Number 70, Theatrum Orbis Terrarum, Amsterdam and New York, 1964.

50. R. Norwood, *Fortification or Architecture Military*, London, 1639. Reprinted in *The English Experience*, Number 545, Theatrum Orbis Terrarum, Amsterdam and New York, 1973.

51. E. Cooke, *The Prospectiue Glasse of Warre, Shewing You a Glimpse of Warre Mystery, In Her Admirable Stratogems*, London, 1628 [STC 5670].

52. I. Blagrave, *A Booke of the Making and Vse of a Staffe, Newly Inuented by the Author, Called the Familiar Staff*, London, 1590 [STC 3118].

53. W. Bourne, *The Arte of Shooting in Great Ordnaunce*, London, 1578. Reprinted in *The English Experience*, Number 117, Theatrum Orbis Terrarum, Amsterdam and New York, 1969.

54. R. Norman, *The safegard of Sailers, or great Rutter, Conning the Courses, Distances, Depthes, Sounding Floudes and Ebbes*, London, 1584. Reprinted in *The English Experience*, Number 827, Theatrum Orbis Terrarum, Amsterdam and New York, 1976.

55. *Mappa Mundi, Otherwise called the Compasse and Cyrcuet of the world*, R. Wyer, London, 1535? [STC 17297].

56. T. Proctor, *A Profitable Worke to this Whole Kingdome. Concerning the mending of all High-ways also for Waters and Iron workes*, London, 1610. Reprinted in *The English Experience*, Number 885, Theatrum Orbis Terrarum, Amsterdam and New York, 1977.

57. W. Gedde, *A Booke of Svndry Dravghtes, Principally serving for Glasiers*, London, 1615. Reprinted in *The English Experience*, Number 316, Theatrum Orbis Terrarum, Amsterdam and New York, 1971.

58. J. B. Merchant, *The Merchants Avizo, Verie Necessary for their Sons and Seruants, when they first send them beyond the seas, as to Spaine and Portingale, or other Countries*, London, 1607. Reprinted in *The English Experience*, Number 98, Theatrum Orbis Terrarum, Amsterdam and New York, 1969.

59. R. More, *The Carpenters Rvle to measure ordinarie Timber*, London, 1602. Reprinted in *The English Experience*, Number 252, Theatrum Orbis Terrarum, Amsterdam and New York, 1970.

60. T. M., *The Silkewormes, and their Flies*, London, 1599 [STC 17994].

61. O. DeSerres, *The Perfect Vse of Silok-Wormes, and their benefit*, London, 1607. Reprinted in *The English Experience*, Number 345, Theatrum Orbis Terrarum, Amsterdam and New York, 1971.

62. W. Stallenge, *Instrvctions for the increasing of Mulberie Trees, and the breeding of Silke-wormes, for the making of Silke in this Kingdome*, London, 1609 [STC 23138].

CHAPTER 3

Format and Page Design in English Renaissance Technical Books: Early Recognition of Reader Context and Literacy Level

If titles of various Renaissance works suggest that books of instructions existed to help readers perform tasks, we may then want to examine these how-to books in terms of issues important in modern technical communication: What did these printed texts look like? Did they differ from other kinds of Renaissance printed books? If so, how did they differ? Since book-trade history tells us that printers and writers provided the kinds of books that would sell, that printers and book-sellers were sensitive to the needs and wants of the reading public, how did writers and printers attempt to write how-to books that would appeal to the perceived audiences for these books?

A quick glance at many of the books mentioned in Chapter 2 suggests that technical books differed in appearance from other kinds of books. Religious works—meditations, homilies, catechisms, books of private devotions, and sermons—were often printed with highly ornate title pages, borders, and illuminated initial block letters. Many were expensive editions, often printed in folio, aimed for sale to the upper classes. Content, printed in English black letter, appeared in dense, undifferentiated text demarcated only by minimal chapter divisions and marginal commentary. Even popular literature and histories, usually appearing in linear, undifferentiated prose, were apparently written to be read thoroughly. In

contrast, technical books (as suggested by many of the illustrations in Chapter 2) frequently demonstrated their writers' or printers' awareness of page design to enhance the accessibility of the text. Examples of these works support the following observations, which will be the focus of this chapter.

1. The organization and page design of many technical how-to books suggest that they were printed for use as reference manuals rather than for thorough, sustained reading. In numerous manuals, material is clearly organized and partitioned, with centered headings in differing sizes and italicized fonts to demarcate divisions of the text. Many, by the middle decades of the sixteenth century, used lists, while many early seventeenth-century books used numbered lists of topics and overviews before chapter headings to describe the content of the chapter. By the mid-sixteenth century, many used marginal descriptive comments or "exdented" headings (headings extending into the page margin) to help readers locate specific ideas.

2. By the third decade of the sixteenth century, most information books included tables of contents as well as indexes, particularly in longer manuals (perhaps sixty or more pages). By the late sixteenth century, white space became a frequently used method of revealing the organization of the text. Some manuals show effective page design and placement of illustrations to allow easy visual access. Longer technical books, such as books on medicinal preparations and surgical methods, first used tables of contents and indexes to show folio locations of specific topics. Pagination by arabic rather than roman numerals would not occur until the latter decades of the sixteenth century.

3. Ramist rhetoric, with its emphasis on displaying information in bracketed tables, directly affected the presentation of technical books, such as medical diagnosis and surgical procedure manuals. The visual power of displaying information in tables and in innovative page design increased sharply after 1570, most likely because of Ramus's influence, a point that will be discussed shortly.

4. The size of many technical books also seems to have been influenced by the needs of users. Most reference manuals were printed as small, compact octavos or quartos that could be easily carried and placed in a pocket or saddlebag. Instructional surgical books can be found that used large folio pages of bracketed instructions—that is, the modern decision tree—to help surgeons in step-by-step procedures during surgery.

5. By the early years of the seventeenth century, roman type had increasingly replaced script type to improve the legibility of print. The merchant or successful craftsman from the upper classes or new readers among the lower classes often read slowly and with difficulty. Efforts to improve the readability of type were often evident in books that displayed improved use of spacing, wider margins, and better paragraph development. As these same books illustrate, typography was becoming an aid to readability.

6. The fact that technical books were apparently written to be used 1) to learn to do a process or 2) to do a task immediately after reading the procedure may

explain why technical books were often more legible than liturgical books, which were expected to be read slowly and thoughtfully. In short, page design in technical books served a functional purpose: the enhanced accessibility of the text for a range of readers, many of whom had low levels of literacy.

7. Advances in technology and advancements in knowledge can be seen in the advances in page design that enabled increasingly complex information to be displayed in visually accessible ways. Thus, the page design of books changed dramatically from 1540 to 1640. Few technical books published after 1620 showed indifference to the importance of page design. Late Renaissance writers and printers seemed aware that page design enhanced the ability of text to contain and then convey meaning.

With these observations in mind, let us turn to representative works that illustrate technical writing and the major principles of format, page design, and organization as these appeared, changed, and moved toward modern format principles during the 1475-1640 period.

EMERGENCE OF MODERN FORMAT IN RENAISSANCE TECHNICAL BOOKS

The Fayt of Armes (1489) [1]

Caxton's translation of Christine DePisan's *The Fayt of Armes* [1], an instruction book about military strategy, is typical of the format used in the first printed books. The book has no title page, but the first two pages announce that the book will be divided into four parts. A descriptive abstract occurs before each chapter in the first part. Throughout *The Fayt*, Caxton began each chapter and section with a forecast statement announcing the organization and signaling the breaks in the text (see Figure 1). In the body of the work, a heading introduced by a ¶ mark announces each chapter (see Figure 2). Although Caxton was not consistent in his use of spacing to present chapter titles and numbers and to demarcate organization, the work was nevertheless tightly organized into four parts. The introduction provides a preview or forecast of the plan of the work. Reflecting the manuscript tradition, the work relies only on ¶ marks to cue readers into shifts in content and movement from one topic to another. That Caxton well understood the importance of organizing text into discernable segments is clearly shown in this early book, perhaps the earliest printed military manual in English.

The Boke of Saint Albans (1486) [2]

A study of format techniques and procedural/process discourse could also justifiably begin with *The Boke of Saint Albans* [2], a beautiful, popular four-part manual of instruction in hawking, hunting, and blazing of arms. The work was written to be used by "gentle men and honest persons" who had an interest in

Here begynneth the table of the rubryshps of the boke of the fayt of armes and of Chyualrye which sayd boke is departyd in to foure partyes/

The fyrst partye deupseth the manere that kynges and prynces oughten to holde in the fayttes of theyr werres and Batayllees after thordre of bokies/dictes/and examples of the most preu & noble conquewours of the world / And how & what maner fayttes ought best to be chosen & the maners that they ought to kepe and holde in theyr offices of armes

Item the second partye speketh after Frontyn of cawtelles & subtyltees of armes which he calleth stratagenies of thordre & manere to fyghte & deffende castellis & cytees after Begece and other auctours/And to make warre & gyue bataylle in ryuers and in the See /

Item the thyrd parte speketh of the dwytes & ryghtes of armes after the lawes & dwyt Wretton

Item the/iiij/partye speketh of the dwytes of armes in the fayttes of saufconduytes/of trewes/of marke/& after of champ of Bataylle/that is of fyghtyng within lystes

Here begynneth the Chapytres of the fyrst boke/

The fyrst chapitre is the prologue/ in which Crystyne excuseth her/to haue dar enterpryse to speke of so hye matere as is contepned in thys sayd boke/   Capitulo / pri°

Item how warres and Batayllees emprysed by iuste and trewe quarell and lade by theyr ryght & dwit/is a thyng of iustyce and suffred of god/   Capitulo /ij

Item how it is not leefful/but only to kynges/ and to souerayn prynces to empryse of theyr finguler auctoryte werres and Batayllees/   Capitulo/iij

Figure 1. Beginning segment of the introduction from
Christine DePisan, *The Fayt of Armes* (1489) [1].

at/as dysperate that they shulde neuer more be of power to
recouere eny good hap or propyer fortune / Wolde for sake
theyre owne cyte/and chese in to some other partyes theyre
place of habytacyon/ But one of theyre pryncys that ryght
wise was q̄ Balpaut kepte hem here fro sayeng that he shuld
fyght ayenst hem yf cas were that they went/ And thus he
putte hem in hope of a better fortune / and assembled theym
ayen al togider/ And of gadred folke he made many kynght
tes/and with such a power as he myght haue he went and
assaylled hanybal that neuere had thoughte that he shulde
haue won soo / and thus toke hym vnputueyed / q̄ was at
that owre so entyerly dyscomfyted that he neuere syth wow
de haue vyctorye vpon the rommaynes /

¶ Here folow eth a short recapptulacyon of som thinges
that ben sayd a fore/   ¶ Capitulo/xxix
TO recapptule shortly almost all the substance of that
whiche Begere wyl saye in hys boke where as by epplo
gacyon in manere of prouerbys at the ende of hyt he sayth
thus/   Thou that wyl haue worshyp in armes / do that
the lore of yougthe kerneth the to be a maystre of the tour
nes q̄ fayctes of knyghthode in thy parfyt age/ For a more
fayre thing it is to say I can thys q̄ that / that to saye ha a
why haue I not kerned/do euere after thy power all that
may bette thyn enemye/and that may be profytable for the
For from that tyme that thou ceasse to greue hym/thou hur
test q̄ lettest thy self/do soo that thou knowe the knyght
tes or euere thou lede hem to the felde/ For moche bettre it is
to doubte hys enemye keppyng hym self vpon hys warde su
rely/than to trust vpon folke that men knowe not in a fel
de/and a grete surete it is for to calle hys enemyes that be
wonne away who that may for they may hurte more sore the

hawking, hunting, and heraldry and learning the appropriate nomenclature for each activity [2, p. A1]. Apparently successful, the book went through extensive revisions and printings, at least twenty-four between 1496 and 1810 [3, pp. 23-24].

Like many early books, both style and organization are clumsy, repetitious, and diffuse—characteristics which suggest that the writer, while generating content, was hearing oral dictation of the text and wrote what he heard. As Ong noted, writing required and produced an organization of thought as it was translated into writing, but this organization was haphazard in the early years of the shift from orality to textuality [4, p. 121]. Thus, early printed works often reflect the aggregative quality of oral discourse instead of the neatly partitioned exposition characteristic of written discourse that appeared in the later seventeenth century. For example, in the section on hawking, remedies for diseases were often mixed with instructions for training. Within the instructions occurs a discussion of proper language and terminology. The style and organization of this book illustrate textualized spoken thought and echo Kellner's description of the natural quality of early English syntax: "that is, it follows much more closely the drift of the ideas, of mental images; the diction, therefore, looks as if it were extemporised, as if written on the spur of the moment" [5, p. 9]. As Ong has also pointed out, early texts, like the spoken thought they echoed, showed some degree of analytic pattern but nothing like that used in technical books of the late Renaissance, where organization and visual presentation became more rigid to help readers see the arrangement of content [4].

Caxton's use of descriptive abstracts and overviews and his crisp demarcation of chapters and parts is remarkable in its anticipation of modern textual practices. Organization on a deeper semantic level is missing, but from a visual perspective, the work illustrates the early English printer's awareness of the importance of visual appeal. It contains the earliest examples of color-printed woodcut illustration in England. These occur in the heraldry section where the shields, used to illustrate the text, appear in black, red, blue, and occasionally gold and are integrated into the text (see Figure 3). Hand coloring with a yellow wash occasionally takes the place of gold printing. Throughout the book, red is used for ¶ marks and for capitals, which serve as the only means of denoting division of material. Descriptive headings, which occur in larger print than the commentary that follows, are set off from the explanatory text by substantial spacing before and after the heading. The result is a visually accessible page in the modern sense of the term. Figure 4 shows a page that defines the parts of coats of arms. Each of the eight parts is defined in a list using parallel phrases introduced by a smaller ¶ mark than that used to denote the major page divisions.

Although different versions of *The Boke of Saint Albans* vary in their use of color, compositors seem to have been attempting to use color to reveal the organization of the text. As was typical of early printed works in England, *The Boke* has no title page, even though a blank page separates each section. All four sections open with an introduction, but only the blazing of arms section is

Off armys barryt crokyt and sharpe as here aft is shewit

Gentill men that be certanli the which bere armis tarad cro
kyt and sharpe as here it apperith in theys armys . and
thay be callid armys barrit for differance of ar;
mys the same man of Wyse palit : and thay be cal;
led crokyt and sharpe . for as it is sayd a fore . ij
colowris ar put to gethyr crokytli and sharpe .
Therfore it shall be sayd that the lorde the which be
ris theys armys berith in this Wyse . first i latyn

Ille portat arma barrata tortuosa et acuta de nigro et auro .
Et gallice sic ꞆIl port barri dauncetee acute de Sable et dor
Anglice sic . ꞆHe berith barris crokyt and sharpe of Sa;
ble and golde .

Now it shall be shewyd of armys that ar bendly barryt .

Ther be forsothe certan armys bendli barrit . and thei be cal
led bendly barrit . and for this cause they be cal de bendly
barrit . for . ij . colouris ar iunyt to gether in eue
ry barre bendly . as it is oppin here i theis armif
And therfore it shall be sayd of him that beris
theis . armys : in this Wyse as folowis . first in la
tyn thus . Ꞇ Ipse portat arma bendata de cu;
bio et auro . Et gallice sic . ꞆIl port barre
bendee de gowblez et dor . Anglice sic . Ꞇ He berith barri
bendy of gowbles and golde .

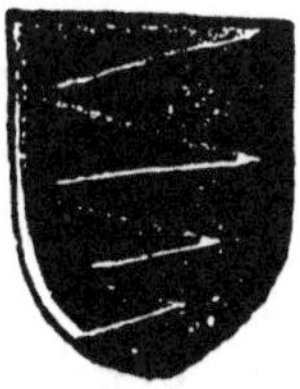

Figure 3. Page with color-printed woodcut illustrations for Blasing of Armes, from D. J. Berners, *The Boke of Saint Albans* (1486) [2].

In so moch that i the fifthe quadrat finiall hit is determyned
of the tokenys of armys . or J proceed to hit : is shewed What
maner of tokeny a gentyll man may Beer .

A gentilman mai not Beer tokynys of armys but of steinig
colowre . that is to say his cotarmure pyntat or ellis J geratt
With pœciouse stonys

Gerattyng þue . tp . bagges of cootarmuris . First With crof
lettis . and of theym ther be . iiij . dyuerse . and thy bene theys
Cros fipyly . Cros paty Cros croflettis . and Cros flory

⊏ The secunde bage is flowre delyœ .
⊏ The thredd baage is roflettys
⊏ The folbrith baage is prymarofe ·
⊏ The fifthe baage is quynfolis .
⊏ The septhe baage is diaclys
⊏ The seuemith baage is chappelettys
⊏ The . viij . baage is Molettys .
⊏ And the . ip . baage is Creffauntis that is to say halfe the
moone . theys be polbœrygis of cootarmuris .

⊏ The fifthe quadrate is calœ Endenfly of . iŋ . diuse Beis
that is to say totally lentally and fpefly .
⊏ Bebally is calœ i armys Whan a cotearmure is calœ En
dentyd of . ŋ . dyuerse colowris m the length of the cotearmure
⊏ Lentalli is calœ m armys Whan yᵉ cotearmure is Endentid
With . ij . dyuerse colowris in the berde of the cotearmure
⊏ Fpefly is called m armys . iŋ . manere Beys Fefy bagy
fefy target and fefy generalt .

Figure 4. Listing of the parts of a coat of arms, from D. J. Berners,
The Boke of Saint Albans (1486) [2].

announced by a title: "Here begynnyth the blasyiong of armys" [2, p. C1]. Like the contemporary introduction, these opening segments state the purpose of the content that follows and the topics to be discussed.

The Vertuouse Boke of Distyllacyon of the Waters of All Maner of Herbes (1527) [6]

The Noble Experyence of the Vertuous Handy Warke of Surgeri (1525) [7]

Books by Hieronymus Von Braunschweig (Jerome of Brunswick) are similar to *The Boke of Saint Albans* in their use of high-quality print, and they are even more illustrative of a format designed for reference. *The Vertuouse Boke of Distyllacyon of the Waters of All Maner of Herbes*, one of the more expensive herbals, is a formidable folio of about 200 pages [6]. It features an ornate title page, a prologue, a two-column index of herbs followed by the signature of the page on which each herb is described, a description section that uses woodcut illustrations of each distillation device used, and a description of the process, with woodcuts integrated at the appropriate point in the description. Following the process-description section is a three-column table—organized according to various human diseases and infirmities, with a reference to the specific item listed under each herb.

Figure 5 shows an excerpt from the index section devoted to diseases of the face. Note column 2, "for the face"; conditions of the face are divided into four categories. The main section of the book features a discussion of each herb—its various uses usually introduced by a rude woodcut of the herb, obviously intended to help the reader identify it in the field. Capital letters, placed before each statement describing the use of the plant, suggest an attempt at listing. Figure 6 shows the specific herb referenced in the index by the item in the list of uses. This book was perhaps the earliest English printed work that used listing as a means of organizing content.

Braunschweig's *The Noble Experyence of the Vertuous Handy Warke of Surgeri* illustrates the same two-column format, a common characteristic of early folios, but the prescription sections are presented in undifferentiated prose [7]. Like *The Boke of Saint Albans*, instructions and description are not visually distinct. The instructions for using tongs to remove arrows, shown in Figure 7, uses only ¶ marks and roman numerals to differentiate steps in the process. With the exception of headings and woodcut drawings, which are integrated into the text at the appropriate point, the page design is rudimentary. Figure 7 contrasts starkly with Figures 15 through 20—a difference which likely can be traced to Ramus's influence on format, a perspective which will be discussed shortly.

To the eares.

Here beginneth the fyfth parte which sheweth all maner of dissases of the eares / and remedyes to the same.

Agaynst defnes.

ii ☽
lxxvi ℛ
cxxii ☦
cxcv ℬ
cxvii ℛ
cclxxv ☿

Agaynst syngyng or pyppynge in the eares.

xvi ℭ
clvii ☽
lxxxvi ℛ
cxxix ℛ
cxcv ℬ

Agaynst payne of the eares.

The fyrst chapytre ℭ
xli ℌ
xlix ☽

Agaynst sores or impostumes in the eares.

ii ☽—
xlv ☿
clxxviii ☿

Agaynst swellynge in the eares

iiii ℱ
xxi ☽
cxcii ☽
ccxii ℭ

For the face.

Here begynneth the sixt parte she wyng the dyseases of the face / & the remedyes for the same.

For to make the face fayre and amyable.

xxvi ℭ
xxxi ℌ
xlv ℬℬ
lxxxvii ℬ
cix ℙ
cxxxix ℌ
cliii ☽
clxii ℭ
clxv ℐ
clxxvi ℭ
ccxx ℭ
cccxvi ℙ
cccxii ℬ

For palenes of the face

xxxii ℌ
lxxvi ℱ
xci ☽
cx ℭ
cxxxvii ℒ
clxviii ℌ
cclxx ℛ

ccxxxvi ℭ

Agaynst the frounces of the face.

ccxcii ℱℱ
ccxciii ℬ

Agaynst the rede pymples in the face.

xxiii ℬ
xlv ℬ
lxxxiii ☽
lxxv ℛ
cxxxiiii ℐ
clxv ℌ
clxviii ℙ
clxxix ℭ

cciiii ℬ
ccvi ℛ
ccvviii ℬ
ccxv ℌ

The water of great burre rotes taken an ou cc and a halfe / and water of rede roses halfe an ounce / and quycke brymstone a dragma myxed to gyder & so let a monethe in the sonne / and the face therwith enoynted .iii. tymes in a day and so let drye by hym selfe is very good for the rose or redenes of the face

Agaynst spottys in the face.

Figure 5. Index to physical conditions, each referenced by chapter and specific use for each plant, from Hieronymus Von Braunschweig, *The Vertuouse Boke of Distyllacyon of the Waters of All Maner of Herbes* (1527) [6].

Dronke in the forsayde maner is good for a body that is wounded and bledeth sore ☽ The same water is good agaynst the hete & reednes of the legges & the blacke blaynes/clowtes or hempe towe wet in the same water and layd theruppon twyse or thryse in a daye / tyll it is slaked ℙ The same water dronke in the mornynge fastynge / and at nyght goynge to bedde/at eche tyme an ounce and a halfe/it is good agaynst the brekinge stone/and the grauell in the lymmes/than shal be kept the bryne in a glas/and in the bothom shall ye fynde a mauer of sande and ye shall se that the stone departeth frome the body ☽ The same water dronke thre tymes in a daye at eche tyme an ounce heleth the gutte in the fondament whā he is trauayled after the laskynge.

¶ Water of polypody. Ca. lxxvi.

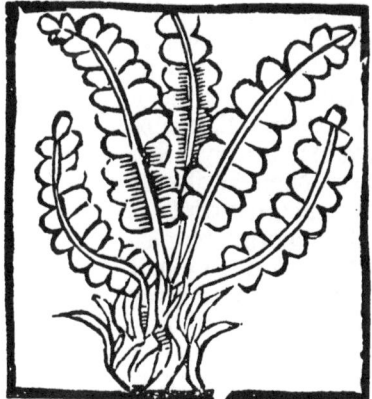

Polipodium in latyn or filica or filix. The best tyme of his dystyllacyon is the rote of them whi che groweth on an oken tree/chopped without the herbe/& distilled be twene bothe our lady dayes ℐ Of the same dronke in the mornyng & at nyght is very good for the cowgh ℬ The same dronke in the forsaid maner is good agaynst madde wyttes & melācolye ℭ The same water dronke in the forsayd maner is good agaynst thought & heuynes of the harte ☽ In a day two or thre tymes dronke of the same/at eche tyme an ounce & a halfe/or two ounces maketh a large brest/ & softeth the bely ☽ Dronke of the same at nyght goyng to bedde/ at eche tyme an ounce & a halfe/ is good for heuy dremes ℱ Dronke of the same in the mornynge & at nyght/ at eche tyme an ounce is good for the yll coloure becaule it clenseth and puryfyed the blode.

¶ Water of fungus. Ca. lxxvii.

Figure 6. Uses of Polypody and Fungus—Early example of listing, from Hieronymus Von Braunschweig, *The Vertuouse Boke of Distyllacyon of the Waters of All Maner of Herbes* (1527) [6].

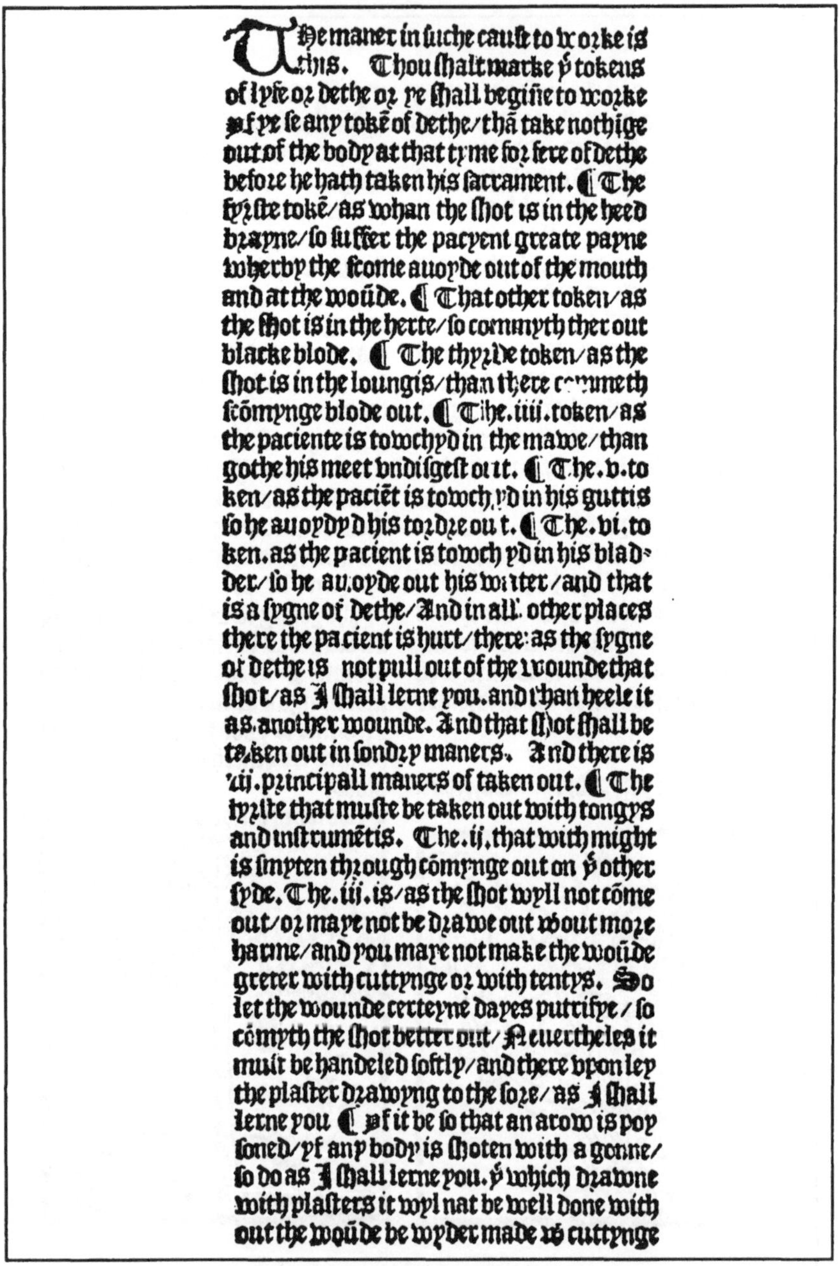

The maner in suche cause to worke is this. Thou shalt marke ꝑ tokens of lyfe oꝛ dethe oꝛ ye shall begiñe to worke ꝫf ye se any tokē of dethe/thã take nothige out of the body at that tyme foꝛ fere of dethe befoꝛe he hath taken his sacrament. ¶ The fyꝛste tokē/as whan the shot is in the heed bꝛayne/so suffer the pacyent greate payne wherby the stome auoyde out of the mouth and at the wouñde. ¶ That other token/as the shot is in the hertte/so cōmmyth ther out blacke blode. ¶ The thyꝛde token/as the shot is in the loungis/than there cōmeth skōmynge blode out. ¶ The. iiii. token/as the paciente is towchyd in the mawe/than gothe his meet vndisgest oꝛ it. ¶ The. b. token/as the paciēt is towch, yd in his guttis so he auoydyd his toꝛdꝛe out. ¶ The. bi. token. as the pacient is towch yd in his blad= der/so he au.oyde out his waꝛter/and that is a sygne of dethe/And in all. other places there the pacient is hurt/there: as the sygne of dethe is not pull out of the wounde that shot/as I shall lerne you. and ꝵhan heele it as. another wounde. And that shot shall be taken out in sondꝛy maners. And there is vij. pꝛincipall maners of taken out. ¶ The fyꝛste that muste be taken out with tongrs and instrumētis. The. ij. that with might is smyten thꝛough cōmynge out on ꝑ other syde. The. iij. is/as the shot wyll not cōme out/oꝛ maye not be dꝛawe out wout moꝛe harme/and you maye not make the wouñde greter with cuttynge oꝛ with tentys. So let the wounde certeyne dayes putrifye/so cōmyth the shot better out/Neuertheles it muit be handeled softly/and there vpon ley the plaster dꝛawyng to the soꝛe/as I shall lerne you ¶ ꝑf it be so that an arow is poy soned/yf any body is shoten with a gonne/ so do as I shall lerne you. ꝑ which dꝛawne with plasters it wyl nat be well done with out the wouñde be wyder made w cuttynge

Figure 7. Early sixteenth century instructions format (for using tongs to remove arrows), from Hieronymus Von Braunschweig, *The Noble Experyence of the Vertuous Handy Warke of Surgeri* (1525) [7].

Early Sixteenth-Century Medical Self-Help Books

Folios, like the two examples just described, generally were printed with a higher quality type than cheaper books such as cookbooks and household management books, which were produced in abundance throughout the English Renaissance. These little books, often published as small duodecimos or quartos that could be easily carried about and used, had broad appeal to readers who recoiled from the cost of larger works. Similarly, books published by Caxton and DeWorde were of higher quality than these same works published by their immediate successors. Technical books, until the latter third of the sixteenth century, were cheaply printed and did not show the high quality reserved for the more expensive folios. For example, Paynell's *A Moche Profitable Treatise Against the Pestilence* (7 editions, 1486-1536) [8] and Thomas Moulton's *This is the Myrour or Glasse of helth* (17 editions, 1531?-1580) [9], two popular self-help medical books, exemplify these cheaply printed first-aid books. Figure 8, from the opening pages of Moulton's book, shows the minimal use of page design and the poor quality of type. Despite their unattractive format, these books were often the only printed source of health education available to middle-class readers.

In all early technical books, word division was not done by syllabication. Printers divided words when they ran out of space at the end of the line. Paragraphing did not occur with any consistency. Until well into the sixteenth century, the ¶ marker was the main means of showing major breaks in thought on the page. As suggested by Caxton's 1489 translation of DePisan's *The Fayt of Armes* illustrated in Figure 2, Caxton was himself insensitive to the value of white space, location of chapter titles, and even spacing, while relying almost totally on the use of the ¶ mark for showing breaks in written discourse. As Ong noted, format would become more important as the importance of language as seen, as opposed to heard, began to control typographic space [4]. In comparing Figures 1 and 8 with later examples, we can see that lack of page design, not just poor quality of type, characterizes early technical books and differentiates them from later Renaissance books.

Hereafter foloweth the Judgemet of all Urynes (1540) [10]

This book, the paradigm for the popular uroscopy books, was first published in 1540. It exhibits the dense prose that characterized most early printed technical books, but it also illustrates the first use of bulleted lists in a printed English book (see Figure 9). Organization and visual display of the content are inconsistent, but many pages are comprised solely of lists—characteristics of urine and what various colors and textures indicate—by using a drawing of a urinal as a bullet. In some editions, as shown in Figure 9, a number appears within each drawing. Note the irregularity of word divisions and the worn quality of the type in this 1553 edition of one of the seventy-two versions of this popular uroscopy.

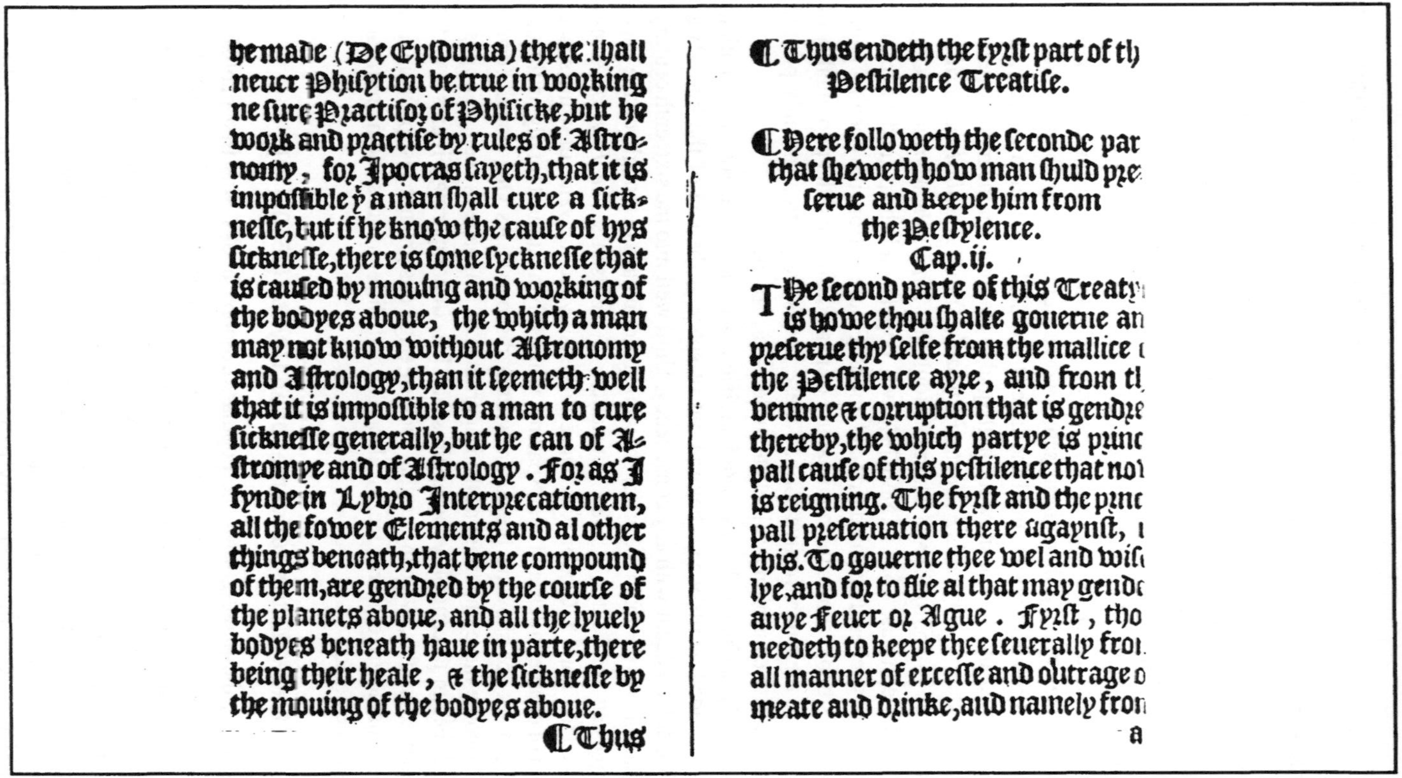

he made (De Epydimia) there shall
neuer Phisytion be true in workyng
ne sure Practisor of Phisicke, but he
work and practise by rules of Astronomy, for Ipocras sayeth, that it is
impossible y a man shall cure a sicknesse, but if he know the cause of hys
sicknesse, there is some sycknesse that
is caused by mouing and working of
the bodyes aboue, the which a man
may not know without Astronomy
and Astrology, than it seemeth well
that it is impossible to a man to cure
sicknesse generally, but he can of A-
stronomye and of Astrology. For as I
fynde in Lybro Interprecationem,
all the fower Elements and al other
things beneath, that bene compound
of them, are gendred by the course of
the planets aboue, and all the lyuely
bodyes beneath haue in parte, there
being their heale, & the sicknesse by
the mouing of the bodyes aboue.
Thus

Thus endeth the fyrst part of th
Pestilence Treatise.

Here followeth the seconde par
that sheweth how man shuld pre
serue and keepe him from
the Pestylence.
Cap. ij.

The second parte of this Treaty
is howe thou shalte gouerne an
preserue thy selfe from the mallice
the Pestilence ayre, and from th
venime & corruption that is gendre
thereby, the which partye is princ
pall cause of this pestilence that no
is reigning. The fyrst and the princ
pall preseruation there agaynst,
this. To gouerne thee wel and wis
lye, and for to flie al that may gende
anye Feuer or Ague. Fyrst, tho
needeth to keepe thee seuerally fro
all manner of excesse and outrage o
meate and drinke, and namely fro
a

Figure 8. Page showing minimal use of format, from Thomas Moulton, *This is the Myrour or Glasse of helth* (1540 edition) [9].

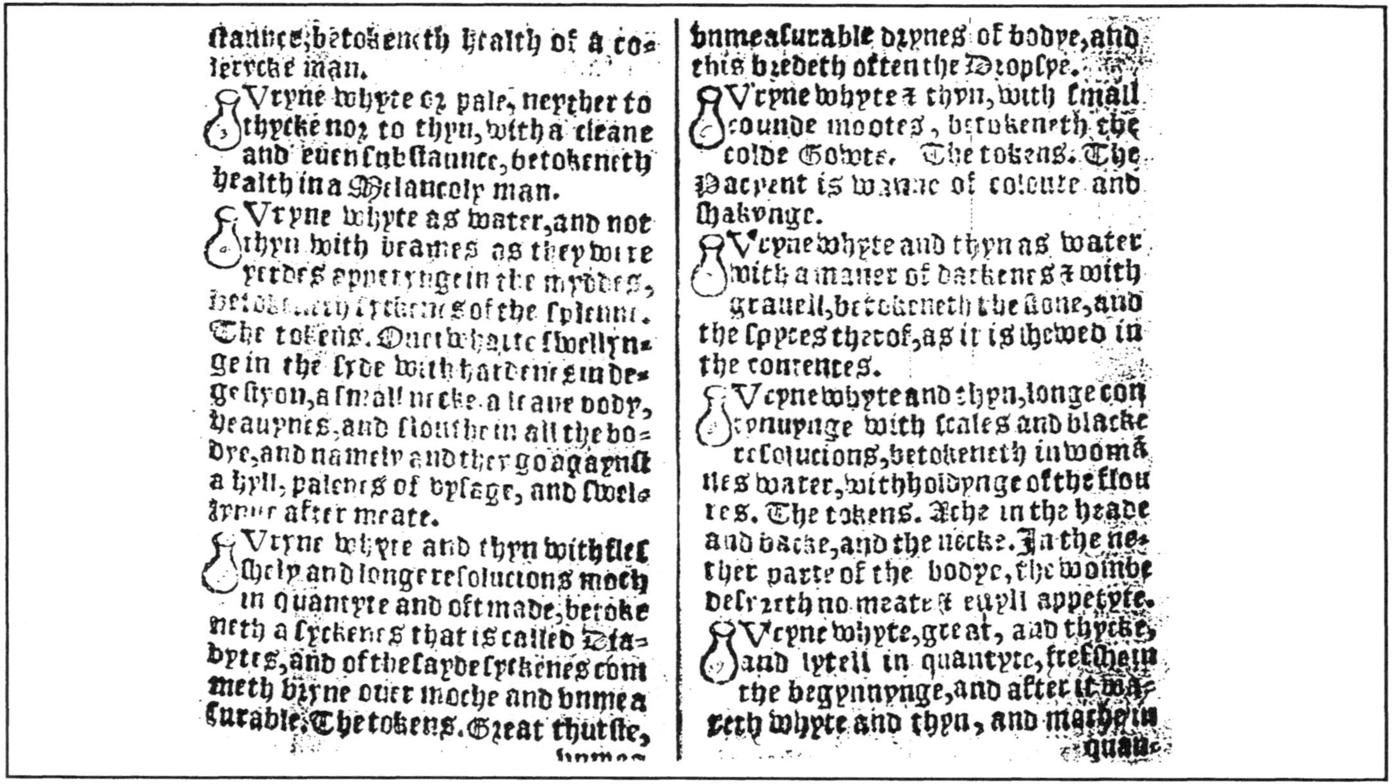

...nature, betokeneth health of a to-
lerable man.

Vrine whyte or pale, neyther to
thycke nor to thyn, with a cleane
and euen substaunce, betokeneth
health in a Melancoly man.

Vrine whyte as water, and not
thyn with braunes as the whyte
Vrine appereyng in the myddes,
betokeneth styrynge of the spliene.
The tokens. Durtle hayte swellynge in the syde with hardenes in de-
gestyon, a small necke, a leane body,
heauynes, and slouth in all the bodye, and namely and ther goo a good
a pyll, palenes of vysage, and suete
feuer after meate.

Vrine whyte and thyn with flex
thyn and longe resolutions moch
in quantyte and oftmade, betokeneth a sycknes that is called Dia-
bytes, and of the sayde sycknes commeth pyne out moche and vnmea
surable. The tokens. Great thurste,

[right column / top]

vnmeasurable drynes of bodye, and
this byddeth often the droplye.

Vrine whyte or thyn, with small
rounde mootes, brokeneth the
colde Gowte. The tokens. The
pacyent is ware of colour and
shakynge.

Vrine whyte and thyn as water
with a maner of darkenes or with
grauell, betokeneth the stone, and
the oppres therof, as it is shewed in
the contentes.

Vrine whyte and thyn, longe con
tynuynge with scales and blacke
resolutions, betokeneth inwarde
vlceratyons, betokeneth of the slou
es water, with holdynge of the flou
res. The tokens. Ache in the heade
and backe, and the necke. In the ne-
ther parte of the bodye, the wombe
desyreth no meate of euyll appetyte.

Vrine whyte, great, and thycke
and lytell in quantyte, betokeneth
the begynnynge, and after commeth
Vrine whyte, thyn, and meane in
quan...

49

Figure 9.   First use of bullets, from D. Smyth, *Hereafter foloweth the Judgemet of all Urynes* (1553 edition) [10].

The Breuiary of Helthe (1547) [11]

The Breuiary of Helthe by Andrew Borde, one of the most prominent English physicians of the sixteenth century, exhibits attempts to use page design to reveal content—centered headings, italics, varying type size—but typography is used excessively and irregularly [11]. While marginal commentary was first used in books published in the 1490s, *The Breuiary* was one of the first technical books to use marginal headings, abbreviated versions of the main headings, to make the topics on each page accessible. As Figure 10 shows, Borde used centered headings, announced by ¶ marks. The headings usually occur in a smaller type than the commentary, although inconsistencies in type size occur. The work opens with a substantial introduction followed by an alphabetized presentation of diseases and conditions. Discussion of each affliction proceeds in the same way: chapter number and title, etymology of the word and its rendering in English, causes of the condition, and remedies. *The Breuiary* consisted of two books. The table, located at the end of The First Boke, compiled the chapter numbers and alphabetized afflictions and noted their location in the book by folio number. The Second Boke, approximately one-third the length of the The First Boke, used the same organization and format.

Bulleins Bulwarke of Defece against all Sicknes, Sornes, and Woundes, that Dooe Daily Affault Mankinde (1562) [12]

This substantial folio of over 200 pages illustrates marked improvements in page design over *The Breuiary of Helthe*: consistent use of spacing, centered headings, and use of italics, and higher quality woodcuts than those used in Braunschweig's books (see Figure 11 and compare it with Figure 6). The use of exdented headings and marginal commentary appeared in the late fifteenth century in devotional works and in about 1550 in technical books. *Bulleins Bulwarke* shows an early example of marginal glosses and a technique now commonly advocated in instructional writing: organizing content around questions that, in turn, become headings which anticipate and answer questions the reader might ask about the material [12]. These questions and answers appear on the page in the order in which the reader might ask them (see Figure 11). The work is organized about dialogues between Marcellus and Hillarius, Sorrenes and Chyrurgi, Sicknes and Health. The marginal commentary and dialogic structure reveal the continued presence of the oral instructional tradition in technical texts and the attempts by writers to capture in text the essence of oral interchange during instruction.

PETER RAMUS'S INFLUENCE ON PAGE DESIGN

While the development of typography, the demand for books, and the spread of literacy definitely worked to improve the ways of presenting information, Ramist dialectic was likely the single greatest philosophical influence on changes in the

The Breuyary

thargos, Memoria, and Cancer.

❡ For Onix loke in the capytle named Piofis.

❡ For Cyfophagos loke in the capytle named Yfophagus.

🖝 The .264. Capytle doth ſhetwe of an blcer in the noſe.

Ulcer. OZenai is the grcke wo?de. Jn latyn it is named Vlcera narium. Jn englyſhe it is named an blcer o? ſcre in the noſe.

❡ The cauſe of this impediment.

❡ This impediment dothe come of a fylthy and an eupl humour the which doth come frō the b?ain and heed ingend?ed of rcume and co?rupte bloce.

❡ A remedy.

❡ Jn this matter rcume muſt be purged, as it doth appere in the capitle named rcume, than beware prcke nat the noſe no? tuche it nat ercepte b?gent cauſes cauſeth the contrary. And bſe gargariccs & ſternutacions, J wyll counſel no man to bſe behement o? er?reme ſternutacions fo? perturbatyng the b?ayne, gentyl ſternutacions is bſed after this fo?t, fy?ſte a man ryſinge from ſlepe o? compnge ſodenly out of a houſe and lokyng into the clymēt o? ſonne ſhall neſe twiſe o? th?iſe o? cls put a ſtrawe o? a ryſſhe into the noſe & tyckle the ryſſhe o? the ſtrawe in the noſe and it wyll make ſternutacions, the pouder of peper the pouder of clibo?us albus ſnuft o? blowē into the noſe dothe make quycke ſternutacions, but in this matter J do aduertiſe cuery man nat to take to much of theſe pouders at a tyme fo? troblynge the ſeccude p?incipal mem= ber which is the b?ayne, and they the wbiche wyll uat neſe ſoppe the noſeth?ylles with the fo?e fynger and the thome bpon the noſe and nat within the noſeth?ylles, and if they wolde they can nat neſe al maner of medccines natwithſtan dynge, howe be it J wolde counſell almen takynge a thing to p?ouoke ſuche matters to make no reſtryctions.

❡ Thus endeth the letter of D. And here fo= loweth the letter of P.

The

Figure 10. Use of centered headings, from Andrew Borde, *The Breuiary of Helthe* (1547) [11].

The booke of Simples.

Marcellus.

What saie you of bones?

Hillarius.

Bones is the timber and strong postes, which coupleth the bodie of euery liuyng man, beaste, fisshe, and foule together: without which we might not be perfite. If any of them are broken, then are we lame. Thus bones bee not onely good to nature, when we are liuyng: but *ossa humana*, mannes bones, are good in medicen. As example,

The bones of {
Man beaten into pouder, doe greatly drie moist humours and sores. &c.

Lions strongly smitten together, will bryng light fire, and are moste drie of nature.

Dogges beaten into pouder, and drunke against the fallyng sickenesse.

Hennes burnt with Egshelles, made in pouder, maketh a frettyng pouder.

Pigges are good for wrytyng Tables, and to kille wormes in the stomake, dronke in wormewood wine.

Many sea fisshes ar holsome in medicen, against poison *Sepia*, or Cuttle, colde and drie, wil cleanse the skinne, and the iyen.
}

Marcellus.

What saie you of vrine?

Hillarius.

It is the whaie of the blood, conuaighed by the raines, into the bladder: hot and drie of nature, verie salt, accordyng to the complexion of the bodie that make it, as man and beast. Foule pisse not, for the moister is tourned into feathers. *Aristotle de natura animalibuns* Vrine is vsed in medicen. As example.

The vrin of {
Man is weakest, except gelded swine, and is wholsome to be drunke against venime.

Younge bote, to be distilled with Letarges of metal, a water to wash against Leprosie, is made therof.

Mule, to wasshe handes and feete, and againste the goute, and paines of the iointes.

Goates or Camelles, to be dronke againste dropsie, or swellyng of the bellie.

Bore, with oile of wormewood, anointe the bellie, and kill wormes.

Dogges will kill wartes, and ringwormes.

Vrine, will corode and frette, clense and skower corrupted humers.
}

Marcellus.

what

Figure 11. Text in dialogic form, using brackets to group information, from W. Bulleins, *Bulleins Bulwarke of Defece against all Sicknes, Sornes, and Woundes, that Dooe Daily Affault Mankinde* (1562) [12].

visual presentation of information. Ramus's logic, although it was replaced by Cartesian logic in the latter seventeenth century, needed less than a century to leave its indelible mark on the consciousness of writers who learned to use the power of the visual display of information as rhetorical strategy. Combined with improvements in print technology, Ramist method provided a significant impetus to major changes in page design. As the preceding examples have suggested, works published after 1560-1580 show a significant design improvement over those published before 1550.

This difference is immediately observable if we compare example page designs from Caxton, Braunschweig, and Moulton (see Figures 2, 6, and 8) with a page from *Bulleins Bulwarke* (Figure 11). Because of the impact of Ramus's method, later sixteenth- and early seventeenth-century writers had a philosophical rationale for enabling readers to "see the text" by employing various page design strategies to enhance visual access. Ramus's impact on page design is thus the most lasting result of his impact on English thought. For that reason, it is important to understand something of the background of Ramist thought.

Ramus's Initial Impact on English Thought

Although Ramus's controversial views on dialectic, rhetoric, and the liberal arts had created a furor within French universities by 1544, Ramist logic did not become an established part of English learning until 1580 [13, pp. 179-180]. However, preparation for Ramus's ideas and logic had begun much earlier. To revitalize Oxford's curriculum, which was languishing from the effects of degenerate scholasticism, Henry VIII decreed in 1530 that Agricola's dialectic be taught along with the works of Aristotle [14, p. 94]. Henry's decree illustrates the reputation of Agricolan logic, which was the foundation for Ramus's own dialectic, and marks the first use of bracketed displays of ideas for dialectic.

Agricola's and ultimately Ramus's logic used a schematic arrangement of logical terms. By dividing or partitioning concepts into increasingly discrete entities and then using brackets to display these dichotomies, Agricola and Ramus created a schematic diagram, a blueprint, of the components of logic. As Figure 12 illustrates, any subject (in this case "dialectic") could be visually divided into its components, then each component further subdivided, until finally, on the right-hand side of the page, the indivisible parts would appear. Logic became a method of "laying things out in a series" [15, p. 125] in a descending order of generality. To Ramus and his followers, logic became a picture of reality, a diagram of the entities (or places, as he called them) that formed reality [14, p. 181]. Roland Macilmaine (M. Roll. Makkylmenaeum) published the first Latin version of Ramus's *Dialecticae* in England and then the first English translation of *Dialecticae* in 1574 [16]. After the English translation of Ramus, his influence on format page design spread rapidly throughout English printed books.

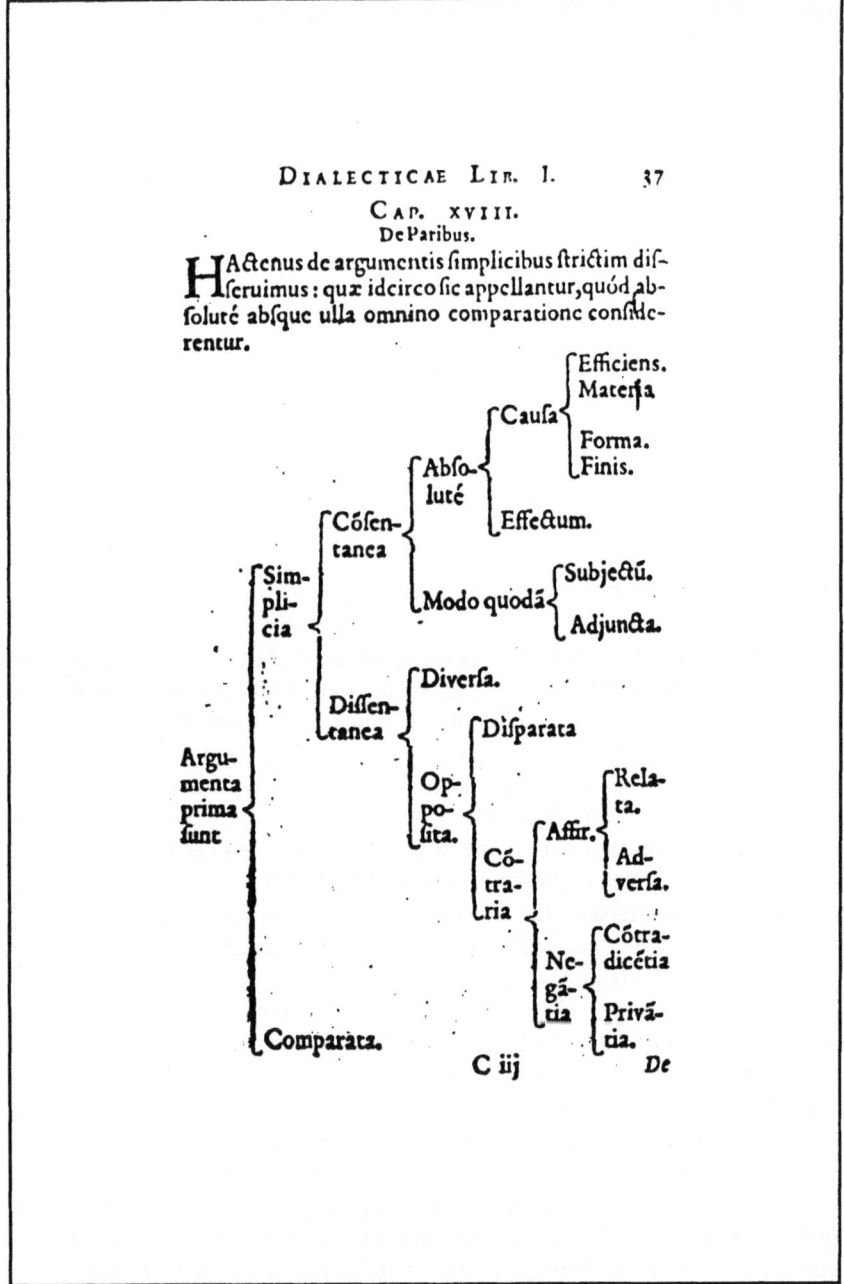

Figure 12. Ramist presentation of the concept of dialectic, from Peter Ramus, *P. Rami Dialecticae Libridvo* (1640 edition) [16].

Ramus's Visual Dialectic

Visual display became an important catalyst in strengthening organization and making the rhetoric of presentation more evident, particularly in practical books. This change can be traced through Ramus's position on existing composition theory. Ramus believed that the medieval logicians' interpretation of Aristotle was confusing and useless. It gave readers, particularly students, no practical understanding of the concepts to which it was applied. With its endless categories and enumeration of disconnected predicables, which existed only for the sake of disputation, logic had lost its practicality in helping students learn, understand, and remember subject matter. At the risk of simplifying, we can state Ramus's goal in this way: to make the truth of any concept accessible and memorable by visual organization based on rigid principles of partition.

Ramus proposed a natural method of organizing and classifying arguments. Because any discipline consists of rules, some of which are more general than others, these rules should be placed in descending order of priority to help the reader's understanding of the relationship among rules. This arrangement by descending order of generality begins with rigid analysis that seeks dichotomies. These dichotomies can then be presented visually by creating informal tables or brackets to group information. Figure 12, from a 1640 edition of Ramus's *Dialecticae* [16] published at Cambridge, shows Ramus's visualized divisions of dialectic.

Howell interpreted Ramus's version of dialectic as beginning with a definition:

> of that whiche is to be handled, must be first set downe, and then the diuision of the same into the members, and the generall properties of the same, and then the diuerse sorts of it, if there be anie; so proceeding vntill by fitte and apte passages or transitions, the whole be so farre handled, that it can be no more deuiced [quoted in 13, p. 221].

Material from any discipline could be "sown asunder" by this kind of logic. Ramus argued that this method was not only simple and clear but also objectively true, that the content of every science falls of itself into dichotomies or natural partitions, and, according to Perry Miller, "that all disciplines can be diagrammed in a chart of successive foliations" [15, p. 127]. According to Miller,

> Dichotomy thus enables the student to "see Gods Logick in things." The work of the intellect, as trained by the Ramist logic, was primarily to perceive and to distinguish, to perceive the divine order and to distinguish its parts by the method of dichotomy [15, p. 128].

Ramists assumed that because One Reason ordered all things, Ramist method allowed the Truth of things to be opened and then compared.

Invention, for Ramus, became synonymous with arrangement, finding the appropriate dichotomies or partitions within a concept. As Miller commented,

> The essence of the Ramist system was exactly this belief that logic is no more than the distinguishing of entities and the joining of them together, that the function of thinking is primarily discerning and disposing, not investigating or deducing [15, p. 134].

Invention discovers, or lays open to view, dichotomous or clearly partitioned arguments as these unquestionably exist. Invention does not create or devise. Logic derives from nature and the natural mind. God has created the world by creating entities and then placing them in sequences, relations, and patterns that can be visually or spatially displayed [15, pp. 149-150].

Ramism thus anticipated current views held by technical communicators that format and organization should reveal the content of the text. In addition, Ramism anticipated the objectivist view of reality—that the world exists. By using our minds we can know it, and by language we can code reality correctly because our minds are part of God's mind. As a result, we *can* distinguish truth from falsehood. Finding truth becomes a process of classification and exploration instead of assertion and proof. To minds tired of a degenerate scholasticism that extracted one proposition from another and snarled over deduction with no goal beyond disputation for the sake of disputation, Ramus's method offered refreshing relief. Truth could be efficiently displayed, visualized, understood, and remembered. Ramist method became the rational justification for procedure and practice because it was a logic derived from the natural processes of the mind. Ramists argued that decency and order prevailed both in the mind and among things.

WORKS SHOWING THE INFLUENCE OF RAMUS ON PAGE DESIGN

The Castle of Helth (1539) [17]

That technical writers should be drawn to Ramist views can be seen in *The Castle of Helth,* first published by Sir Thomas Elyot in 1539 [17]. In fact, the widespread popularity of Elyot's *Castle* was perhaps due to its use of Agricolan-Ramist method for presenting content in an efficient, visually accessible way. Elyot's little manual, small enough to be easily carried about, was the first English technical manual to use a bracketed and tabular page design to present medical conditions and their causes (see Figure 13). *The Castle of Helth* did not exemplify strict Ramist method. However, evidence suggests that Elyot apparently read Agricola and sustained his connection with Oxford and Henry VIII during the time Agricolan logic was incorporated into Oxford's curriculum (1530), approximately four years before he wrote *The Castle* (1534-1535). In addition, Elyot, known for his thoughtful assimilation of humanist ideas, was associated with some of the best and most progressive English thinkers through his friendship with Sir Thomas More. Thus, the design of even the first edition shows that Elyot

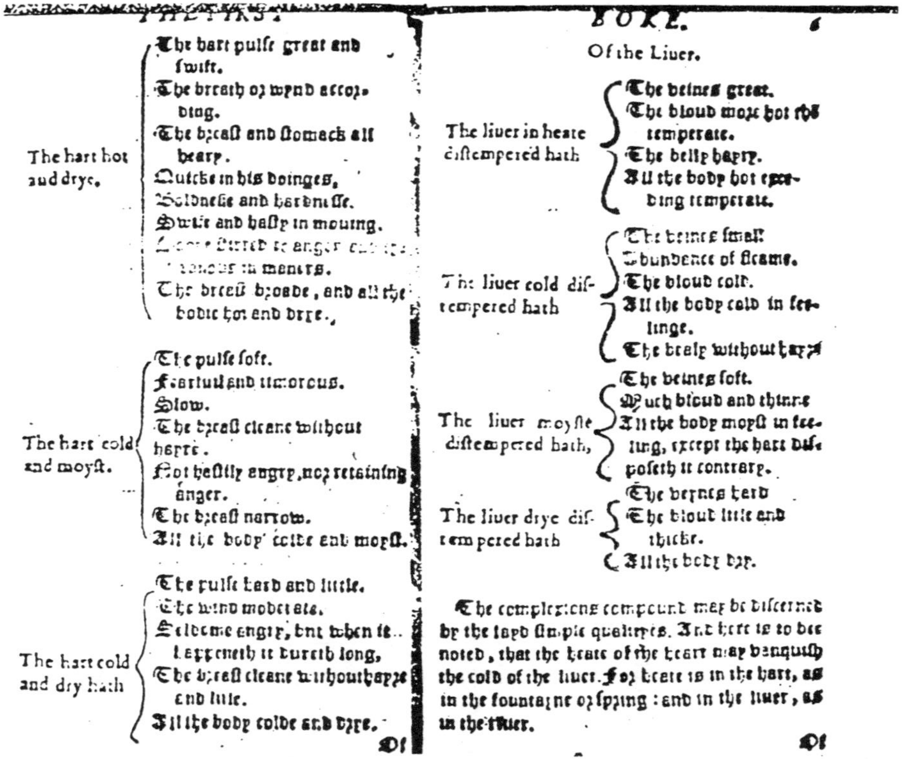

THE FIRST

The hart hot and drye.
The hart pulse great and swift.
The breath or wynd accordyng.
The brest and stomack all heary.
Quicke in his doinges,
Boldnesse and hardinesse.
Swift and hasty in mouing.
... stirred to anger ... tyrannous in maners.
The brest broade, and all the bodie hot and drye.

The hart cold and moyst.
The pulse soft.
Fearfull and timorous.
Slow.
The brest cleane without heere.
Not hastily angry, nor retaining anger.
The brest narrow.
All the body colde and moyst.

The hart cold and dry hath
The pulse hard and little.
The wind moderate.
Seldome angry, but when it happeneth it dureth long,
The brest cleane without heare and litle.
All the body colde and drye.

Of

BOKE. 6

Of the Liuer.

The liuer in heate distempered hath
The veines great.
The bloud more hot than temperate.
The belly heary.
All the body hot excedyng temperate.

The liuer cold distempered hath
The veines small
Abundance of fleume.
The bloud cold.
All the body cold in felinge.
The bely without heare

The liuer moyste distempered hath,
The veines soft.
Much bloud and thinne
All the body moyst in felyng, except the hart disposeth it contrary.

The liuer drye distempered hath
The veines hard
The bloud litle and thicke.
All the body dry.

The complexions compound may be discerned by the sayd simple qualities. And here is to bee noted, that the heate of the heart may vanquish the cold of the liuer. For heate is in the hart, as in the fountayne or spring: and in the liuer, as in the riuer.

Of

Figure 13. First use of brackets to classify medical conditions, from Thomas Elyot, *The Castle of Helth* (1539 edition) [17].

wanted his work to be quickly visually accessible and that he considered visual accessibility a rhetorical factor compatible with his effort to use non-latinate syntax in writing *The Boke Named the Gouernour* [18].

The use of bracketed pages in Part One of *The Castle of Helth* (Figure 13) places the work in stark contrast to other self-help medical manuals, such as Moulton's (see Figure 8), which were written before 1550. That *Bulleins Bulwarke* [12] was also likely influenced by Ramist thought can be seen in Figure 11. This page, with its use of brackets, echos the use of bracketed displays by Elyot in *The Castle* (Figure 13) and became a model for later works that experimented with the use of brackets for revealing content and organization.

The concise, deliberate visual design of *The Castle of Helth* becomes even more evident when we compare it with *The Boke Named the Gouernour* [18], which is characterized by dense, undifferentiated discourse and minimal partition among major segments. While *The Castle* was designed to be used for quick reference, the *Gouernour* demands attentive reading to absorb its aureate syntax, about which more will be said in Chapter 5.

Ramist Method and the Technological Impetus Toward Clarity in Medical Books

The interplay between philosophical forces and evolving technology of type can be observed in works that appeared after 1550-1560. In short, advances in rhetorical perspective and in print technology provided a range of new and powerful tools for designing pages for easy visual accessibility. Writers and printers could and did select a variety of tabular displays, brackets, font sizes, centered heads, and marginal descriptors and headings to reveal the content and organization of the text. With Ramist method, truth became associated with visual clarity. As Ong observed, Ramist thought and typography "heightened the value of the visual imagination and the visual memory over the auditory imagination and the auditory memory" [14, p. 167]. The result was a change in sensibility driven by a philosophy approving the use of typography to diagram ideas.

The pervasiveness of Ramus's influence in this one area of page design is clear when we see that, after 1550, many books giving instructions for prescriptions used bracketed displays to classify and list ingredients. Two notable examples were John Banister's *A Needefull, new, and necessarie treatise of Chyrurgerie . . .* (1575) [19] and William Clowes's *A Profitable and Necessarie Booke of Obseruations . . .* (1596) [20], both of which were written for physicians. Figure 14 shows the use of extended headings, boldface type, listing, and artful spatial arrangement to organize content in Clowes's book.

Before examining other works that illustrate the Renaissance evolution in typography and page design, we should examine Ramism as it affected the visual design of a number of major medical books, perhaps the largest category of technical writing in the English Renaissance.

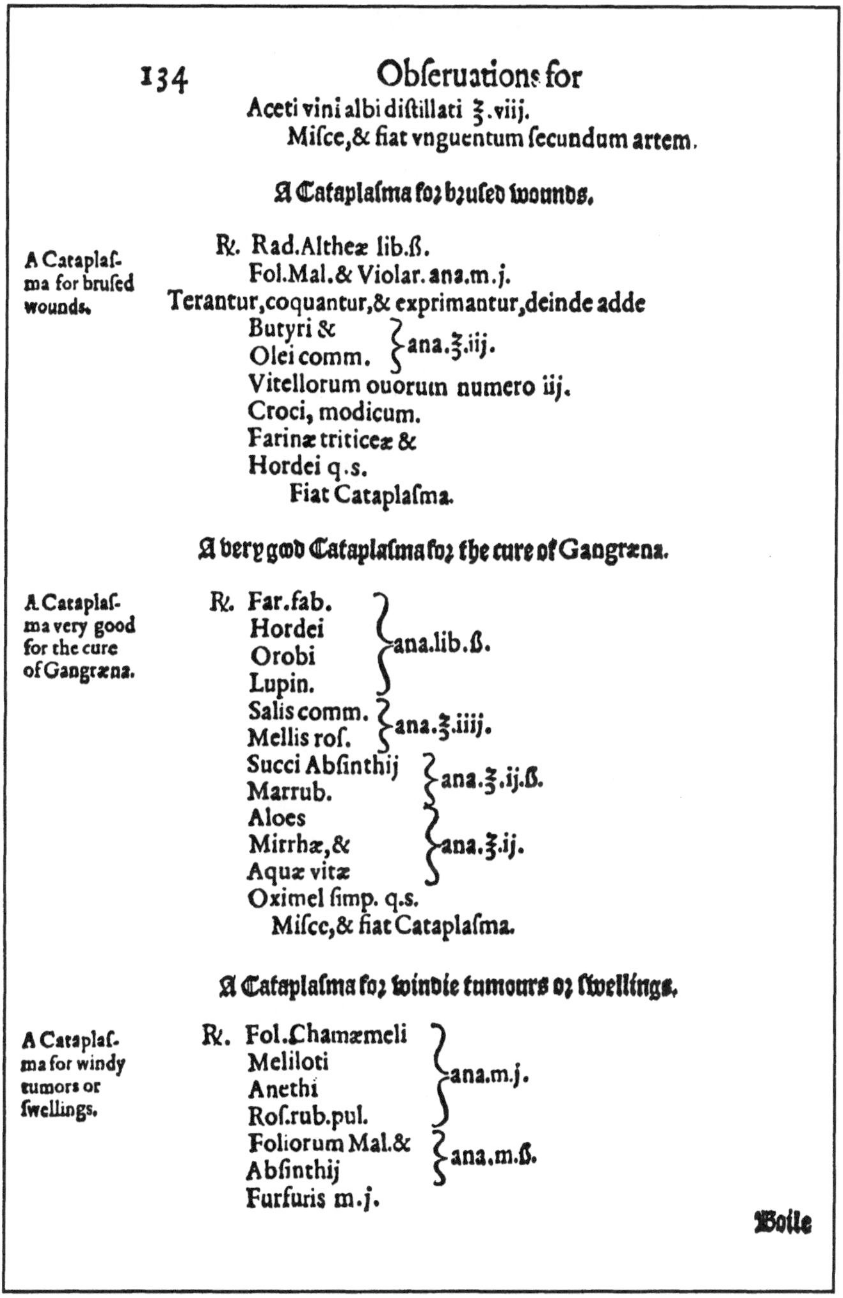

134   Obſeruations for

Aceti vini albi diſtillati ℥.viij.

Miſce,& fiat vnguentum ſecundum artem.

A Cataplaſma foʒ bʒuſed woundſ.

A Cataplaſ-
ma for bruſed
woundſ.

℞. Rad.Altheæ lib.ß.

Fol.Mal.& Violar. ana.m.j.

Terantur,coquantur,& exprimantur,deinde adde

Butyri &
Olei comm. } ana.℥.iij.

Vitellorum ouorum numero iij.

Croci, modicum.

Farinæ triticeæ &

Hordei q.s.

Fiat Cataplaſma.

A very good Cataplaſma foʒ the cure of Gangræna.

A Cataplaſ-
ma very good
for the cure
of Gangræna.

℞. Far.fab.
Hordei
Orobi
Lupin.   } ana.lib.ß.

Salis comm.
Mellis roſ.   } ana.℥.iiij.

Succi Abſinthij
Marrub.   } ana.℥.ij.ß.

Aloes
Mirrhæ,&
Aquæ vitæ   } ana.℥.ij.

Oximel ſimp. q.s.

Miſce,& fiat Cataplaſma.

A Cataplaſma foʒ windie tumourſ oʒ ſwellingſ.

A Cataplaſ-
ma for windy
tumors or
ſwellings.

℞. Fol.Chamæmeli
Meliloti
Anethi
Roſ.rub.pul.   } ana.m.j.

Foliorum Mal.&
Abſinthij   } ana.m.ß.

Furſuris m.j.

Bolle

Figure 14. Listing of medicinal instructions, from William Clowes,
A Profitable and Necessarie Booke of Obseruations . . . (1596) [20].

Certaine workes of chirurgerie (1563) [21]

Thomas Gale, a prominent English surgeon trained on the continent, wrote perhaps the first medical book to clearly reflect the influence of Ramus's logic: *Certaine workes of chirurgerie* [21]. While the book as a whole is dense, undifferentiated prose without marginal commentary, the work includes three Ramist tables—one a standard-size page and two fold-outs—that define by classification wounds, fractures, and ulcers. Figure 15 shows the Ramist table that defines by type the differences among wounds. The second portion of Gale's book emphasizes pharmaceutical preparations and, like works by Banister and Clowes, uses listings and brackets to display ingredients and their quantities. In less than forty years, page design, as a result of Ramist thought, had undergone a major transformation. While Ramist logic would cease to be a factor in either rhetoric or argument by the close of the seventeenth century, Ramist theory, drawing on improvements in typography, was destined to leave its distinctive mark on page design techniques in printed books.

A Worthy Treatise of the eyes (1587) [22]

A Needefull, new, and necessarie treatise of Chyrurgerie (1575) [19]

Visual display of information is irrevocably bound to and dependent on organization of content. Ramus, anticipating much current technical communication research, advocated hierarchies of information—either in bracketed displays or in clearly demarcated linear discourse. As Macilmaine (Makkylmenaeum) stated in his English translation of *Dialecticae*, Ramus's method lifted Aristotelian logic from the confusing method of scholastic argument:

> The methode [of Ramus] is a disposition by the which amonge many proposition of one sort, and by their disposition knowen, that thing which is absolutely most cleare is first placed, and secondly that which is next: and therefore it contynually procedethe fromtyhe most generall to the speciall and singuler. By this methode we proceade from the antecedent more absolutely knowen to proue the consequent, which is not so manifestly knowen: this is the only methode with Aristotle did obserue [23, p. 94].

This arrangement by descending order of generality was immediately grasped by medical writers and applied to a variety of medical manuals. Macilmaine provided a detailed example of how Ramist method could be applied to medicine:

> Yf thou be a Phistion and willing to teache (as for example), of a feuer, this methode willethe thee to shewe first the definition, that is, what a feuer is, next the deuision, declaring what sorte of feuer it is, next to the deuisio . . . : thirdly to come to the places of inuention, and shewe fyrst the causes of the feuer euery one in order. . . . The second place is the effecte, shewe then what the

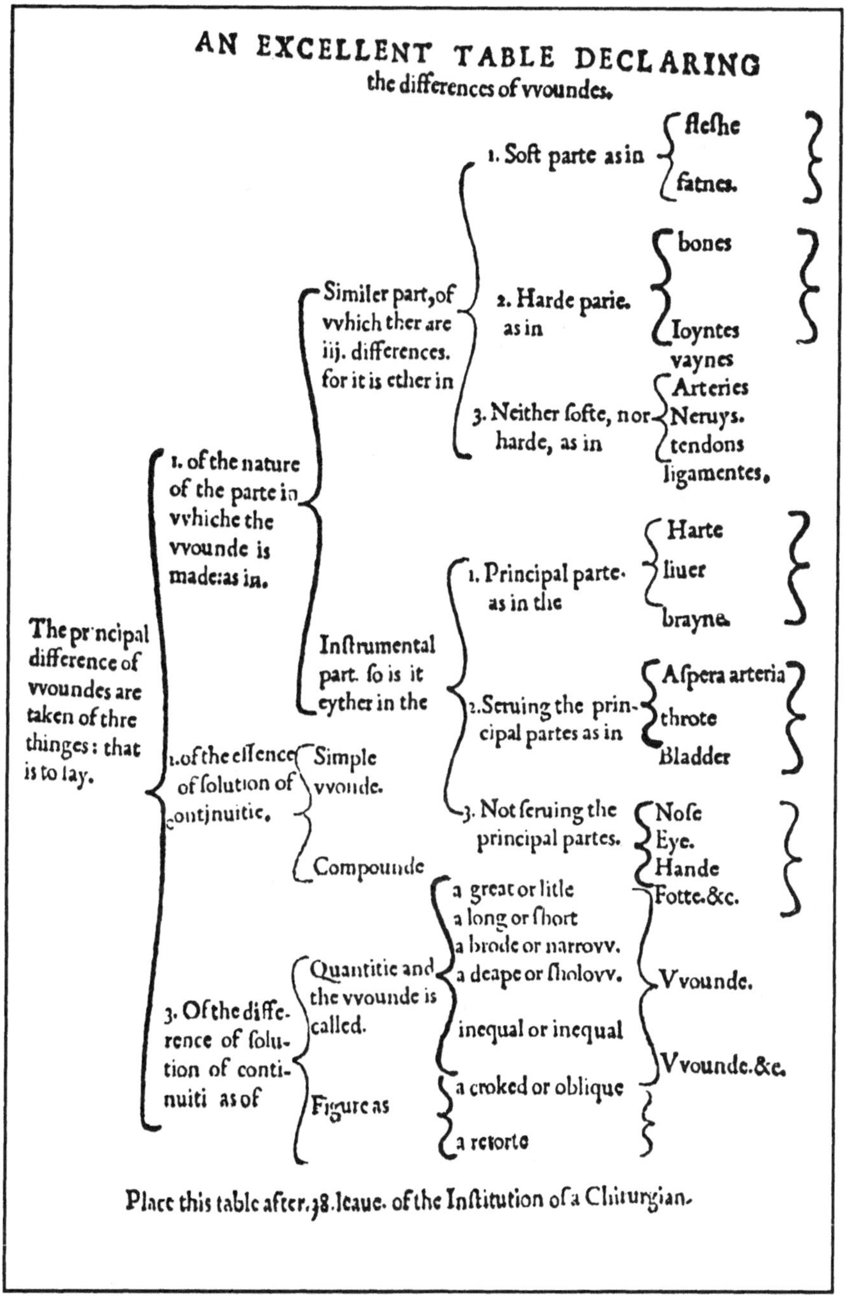

AN EXCELLENT TABLE DECLARING
the differences of vvoundes.

The principal difference of vvoundes are taken of thre thinges: that is to say.

1. of the nature of the parte in vvhiche the vvounde is made: as in.

Similer part, of vvhich ther are iij. differences. for it is ether in

1. Soft parte as in — fleshe / fatnes.

2. Harde parie. as in — bones / Ioyntes vaynes

3. Neither softe, nor harde, as in — Arteries Neruys. tendons ligamentes,

2. of the essence of solution of continuitie,

Instrumental part. so is it eyther in the

1. Principal parte. as in the — Harte liuer brayne.

2. Seruing the principal partes as in — Aspera arteria throte Bladder

3. Not seruing the principal partes. — Nose Eye. Hande Fotte. &c.

Simple vvounde.

Compounde

3. Of the difference of solution of continuiti as of

Quantitie and the vvounde is called.

a great or litle
a long or short
a brode or narrovv.
a deape or sholovv.

inequal or inequal — Vvounde.

Figure as

a croked or oblique — Vvounde. &c.

a retorte

Place this table after. 38. leaue. of the Institution of a Chirurgian.

Figure 15. Ramist table for summarizing kinds of wounds, from Thomas Gale, *Certaine workes of chirurgerie* (1563) [21].

feuer is able to bring forth. . . . The third place wishethe thee to tell the subjecte of the feuer, whether it be in the vaines, artiers, or els where. The fowrthe is to shewe the signes and tokens which appeare to pretende lyfe or death. . . . And last come to the confirmyng of thy sayinges by examples, autyhorities, and . . . by histories and long experience [23, pp. 13-14].

How this laying out of information was done in linear prose can be seen in Jacques Guillemeau's *A Worthy Treatise of the eyes* (1587) [22], which followed the Ramist method described by Macilmaine. For example, in his discussion of how to recognize and treat various problems, Guillemeau structured his description and analysis of each condition in typical Ramist fashion that moves from the general to the particular: definition, causes, and cures.

In his work on the treatment of ulcers—*A Needefull, new, and necessarie treatise of Chyrurgerie, briefly comprehending the generall and particular curation of Vlcers* (1575) [19]—John Banister structured his discourse as follows: definition of ulcers, differences among ulcers, causes, signs of the presence of ulcers, prognosis, and cure. Each section is announced by a centered heading. At the end of the work, his "Table of Simples" reflects the dichotomous organization characteristic of Ramist presentation (see Chapter 4, Figure 3, p. 100). Banister divided his medicinal compounds into hot and moist, cold and dry, cold and moist, and hot and dry. Each compound was further distinguished by degrees. Much of the material appeared in informal tables introduced by centered heads.

Tables of Svrgerie (1585) [24]

Increased attention to format and page design can be tracked in medical works of other late sixteenth- and early seventeenth-century writers who absorbed Ramist thought. Perhaps the first completely Ramist medical work in English was *Tables of Svrgerie*, described in its subtitle as "Brieflie Comprehending the Whole Art and Practise Thereof In a maruelous good method, collected and gathered out of the best physicians by Hoaratius Morus a Florentine Physician and Faithfvillie Translated out of Latine into our English toonge by Richard Caldwell doctor of physicke" (1585) [24]. The work, consisting of large folio pages, was composed solely of bracketed tables (see Figure 16). Caldwell used Ramist method to set up a way of diagnosing a condition and establishing the correct treatment. That visual clarity was Caldwell's goal for the student reader can be inferred from a comment in the preface, which also suggests that Caldwell expected to be attacked by anti-Ramists: "[I have] contrived in such order, that at one view euerie thing seuerallie might be easilie seene and conceiued, I caused them to bbe set out to the profit of yoong students. I know full well that there will be some which will little regard this our honest paines-taking" [24].

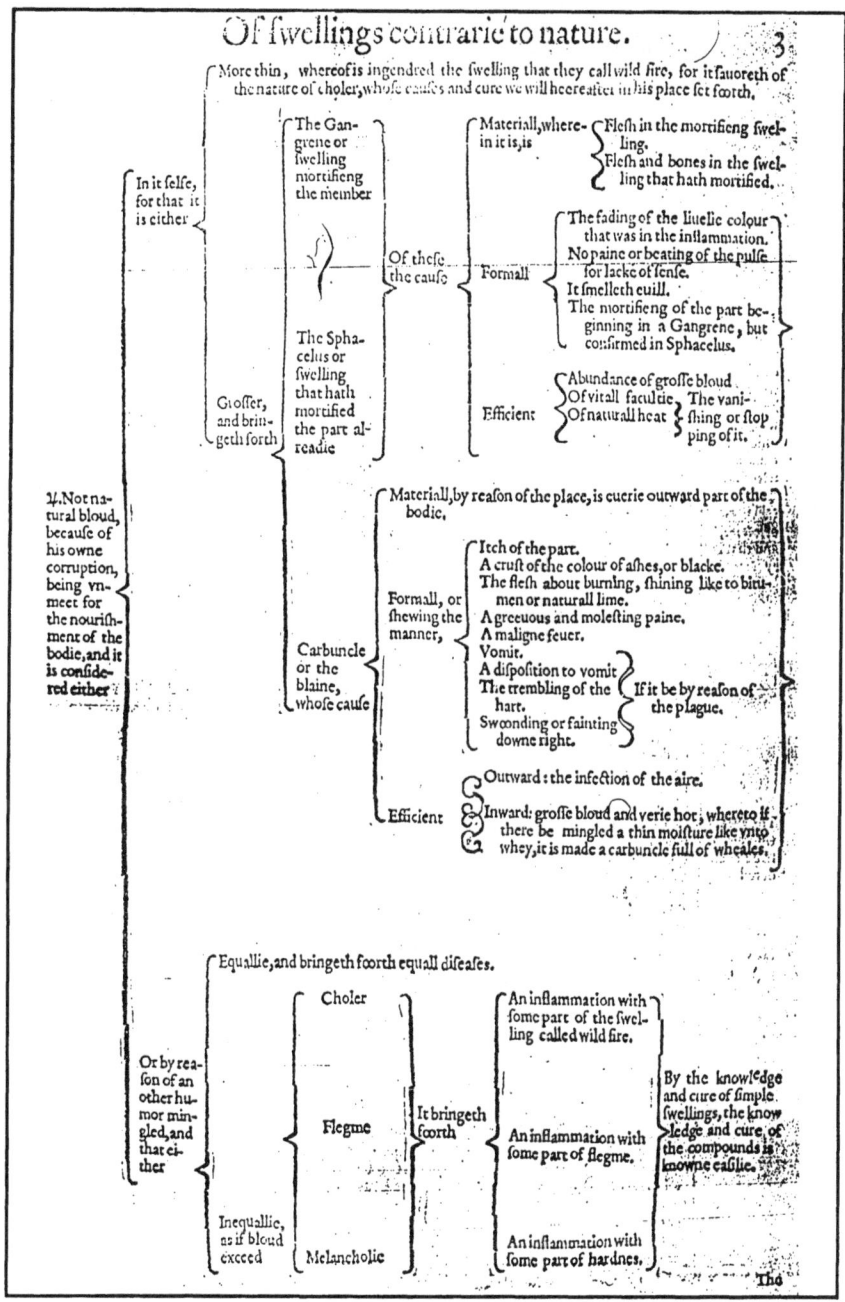

Figure 16. Ramist instructional folio for surgical students, from
Richard Caldwell (translated from H. Morus), *Tables of Svrgerie* (1585) [24].

The chirurgicall lectures of tumors and ulcers delivered in the Chirurgeons Hall (1635) [25]

Ramism is clearly evident in the works of Alexander Read, a well-known English surgeon. Read's *The chirurgicall lectures of tumors and ulcers delivered in the Chirurgeons Hall* (1635) [25] combined discourse developed according to Ramist method with bracketed tables. The lectures opened with a bracketed summary of the contents of the work (see Chapter 4, Figure 2, p. 97), similar to that used in the 1640 Latin version of the *Dialecticae* shown in Figure 12 (p. 54). In *The chirurgicall lectures*, a bracketed summary precedes each discourse section (see Figure 17). Read obviously took great pains to develop a work that would be easy for students to follow during lectures and during individual study. The content of each chapter is immediately evident; and the work as a whole, as well as each individual page, has been designed for efficient reading and memorization.

The analysis of Chyrurgery, Being the Theorique and Practique part therof (1636) [26]

The Cvre of All Sorts of Fevers (1638) [27]

Two additional works for medical students deserve mention because of their thoroughly Ramist visual presentation. Edward Edwards, Doctor of Physic, about whom history has left us little record, wrote two works: *The analysis of Chyrurgery . . .* (1636) [26] and *The Cvre of All Sorts of Fevers* (1638) [27]. The subtitle of this second work immediately reflects, as did Guillemeau's work on eyes, its Ramist purpose: "Both generall, and particular, with their Definition, Kindes, Differences, Causes, Signes, Prognostication, and Manner of Cure." Instead of discourse arranged about each of these main topics, as used by John Banister [19] and Jacques Guillemeau [22], both works by Edwards use bracketed tables exclusively. Figure 18 shows another major characteristic of Ramist works that would influence the design of technical writing: an enumerated summary of ideas preceding the presentation of the ideas so as to reveal the organization of the content that follows.

Edwards's preface on surgery suggests that writers who exhibited a Ramist influence believed that visual organization was directly related to the purpose of the work. Edwards stated that his aim was to enhance the learning of medical students by making medical knowledge easier and quicker to understand and remember:

> I have set forth these briefe instructions, for their better memory: as Load-
> stone to draw them from one degree to another, both in the parts theorick and
> practick, . . . all in analised Tables, in the which if they will diligently bestow
> a little labour, and study therein, I thinke they shall learne more in one month,
> then they shall doe in five, in any other, formerly written in our vulgar tongue
> [26, p. A4].

Reflecting on these Ramist medical writings by Gale, Read, Guillemeau, and Edwards, we can see echoes of Ong's conclusion about Ramist writings in general:

> Paragraphs and centered headings appear, tables are utilized more and more until occasionally whole folio editions are put out with every bit of the text worked piecemeal onto bracketed outlines in dichotomized divisions which show diagrammatically how "specials" are subordinated to "generals" [14, p. 184].

While various forms of brackets and visualized dichotomies and schematics had been used in medieval manuscripts [28, pp. 38-46], after the advent of printing, sixteenth-century writers who were drawn to Ramus's method found the power available in typography an added incentive to experiment with brackets and other ways of creatively revealing content.

Ramus's Influence on Other Kinds of Writing

Ramus's influence can also be found in religious works, particularly Puritan expositions of Scripture, and in geography books, surveying books, and even grammar books. While these books are beyond the scope of my study of Renaissance technical writing, they do show how pervasive Ramus's work was in enhancing page design and enforcing rigid organization on books that do not seem to lend themselves to Ramism. For example, Edward Vaughan's *Ten Introductions* (1594) on how to understand the Bible uses rigid partition, announced by numbered lists, to introduce each part of the exposition [29]. Ramist brackets show where Old Testament stories are located. The same procedure occurred in Valentine Leigh's *The Moste Profitable And commendable Science, or Surveiyng of Landes, Tenementes, and Hereditamentes* (1578) [30].

Some writers, however, attempted to force Ramist method on any and all topics. The result was text with major readability and structural problems. That Ramist brackets were often self-defeating can be seen in Lombarde's *A Preambulation of Kent: Conteining the Description, Hystorie, and Customes of that Shyre* (1598) [31]. Lombarde used page after page of Ramist tables to describe the Shyre, its justice system and its geography. One of his efforts to describe the justice system is shown in Figure 19. Because the Ramist dichotomy is not contained on each page, as in Caldwell's and Edwards's books, the reader loses all sense of the larger organizational concept which the Ramist display is intended to reveal. As Ong observed [14], because of the underlying system of rigid partition that demanded that content be "sown asunder," successive editions of Ramus's works as well as Ramist books would move from dense, undifferentiated prose, first to text with centered heads, exdented heads, and judicious use of white space, then to books that were, like Caldwell's, Lombarde's, and Leigh's, composed almost entirely of tables and bracketed displays that subdued the role of prose as an adequate vehicle

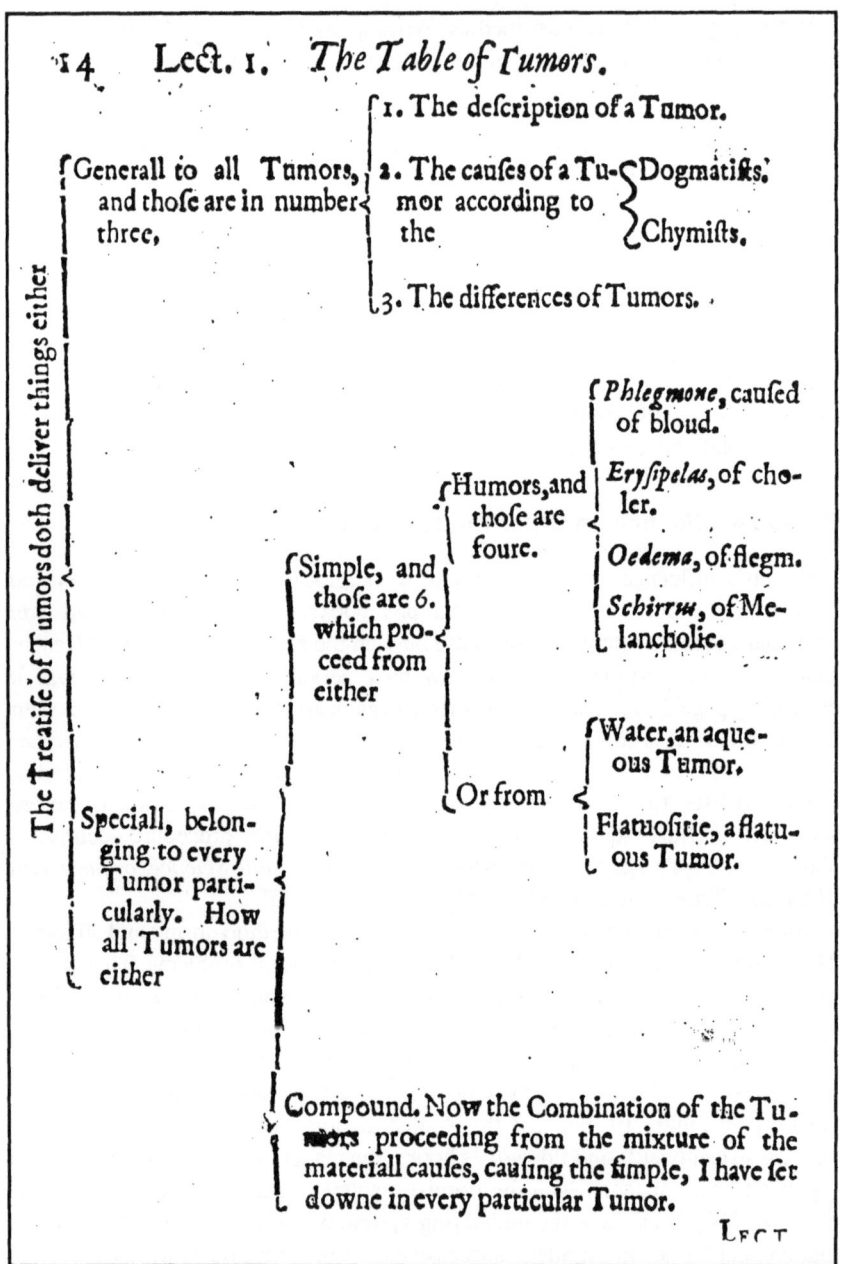

Figure 17. Chapter summary and opening page for Lecture 11, from
Alexander Read, *The chirurgicall lectures of tumors and ulcers
delivered in the Chirurgeons Hall* (1635) [25].

A Treatife of Tumors. Treat. 1. 15

LECT. II.

The generall doctrine of Tumors.

SEeing according to *Ariftotle* in *cap.1.lib.1. phyfic. acro-
af.* in every methodicall tractation, that which is moft
generall ought firft to bee fet downe, then that which is
more fpeciall; becaufe the generall points being fet down,
they exceedingly further the knowledge of the particulars:
I will begin from thofe things which are common to all
fpeciall Tumors, and thofe I make in number three. Firft,
I will declare what a Tumor is. Secondly, which are caufes
of Tumors : Thirdly, which are the maine differences of
Tumors.

Before I come to the effentiall definition of a Tumor, The appellati-
I will fet down the denominations of it. A Tumor in ons of a Tumor.
Greeke is called ὄγκος, that is, a prominence or protube-
rance in the bodie. And from hence the Latine words *uncus*
a crooke or hooke, and *aduncus* bended or crooked, are
derived: Becaufe things bended caufe a fticking out. The
Arabians and barbarous Phyfitians, who follow them, call
all Tumors unnaturall *Apoftemata*, in Latine *abfceffus*: this
word is derived from the greeke verb ἀφίςαϑαι, *abfcedere*:
Becaufe in apoftemes there is a collection of humors in any
place, which have left their own proper feat &c. For humors, which caufe Apoftemes, come from the veines, and
fo leaving ther own naturall receptacle, feat themfelves in
other parts of the body, being dependant and weake. And
Chirurgeans commonly call Tumors wherein there is collection of matter Apoftemes. *Tumor*, which is a Latine
word, and by frequent ufe made familiar in Englifh confe-
rences and difcourfes, is derived from the Latine word
Tumeo, to bee raifed or puffed up : And from *Tumor* commeth *Tumulus* a grave: becaufe it is raifed up higher than
the ground adjacent to it. Now bunchings or ftickings out Differences
of parts of the body, are threefold : for either they are of Tumors
naturall, and then they ferve for the comelineffe of the the body.

Figure 17. (Cont'd.)

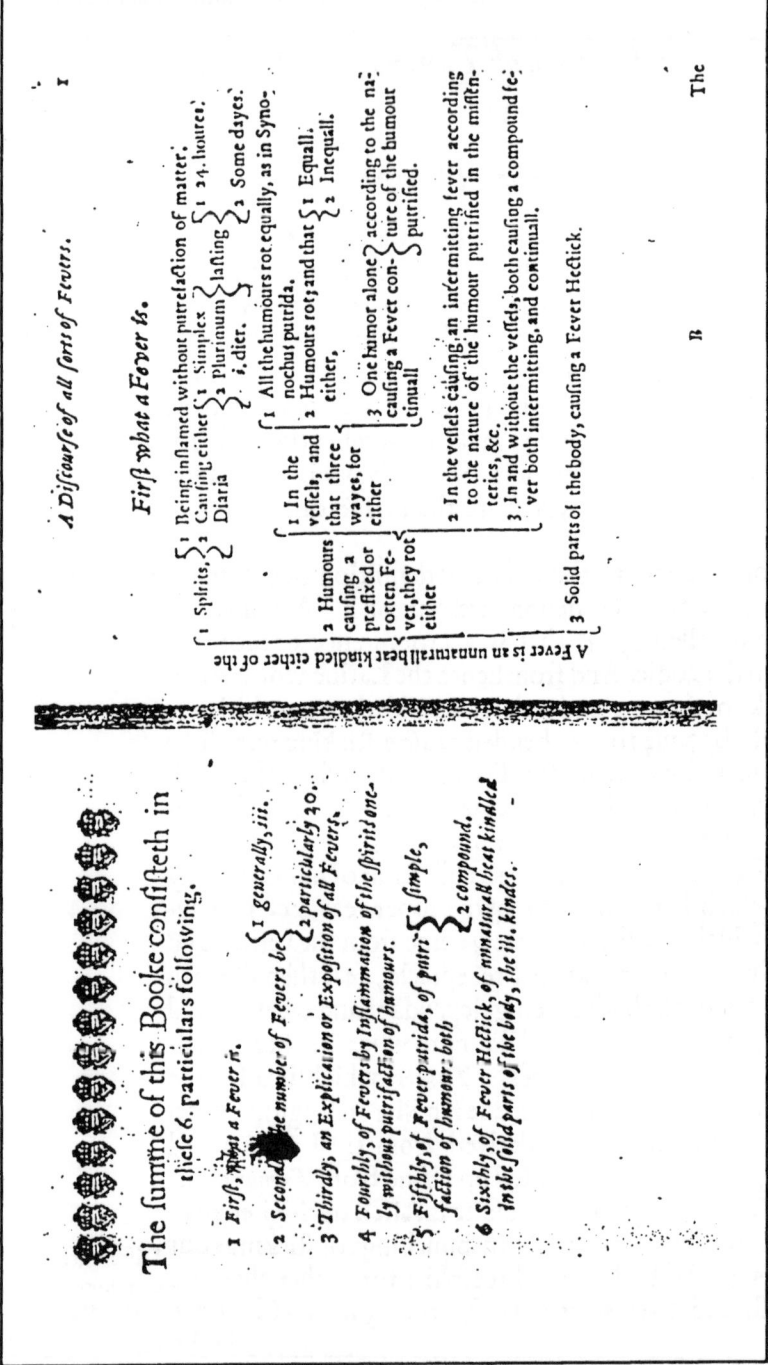

Figure 18. Chapter summary and opening page from Edward Edwards, *The Cvre of All Sorts of Fevers* (1638) [27].

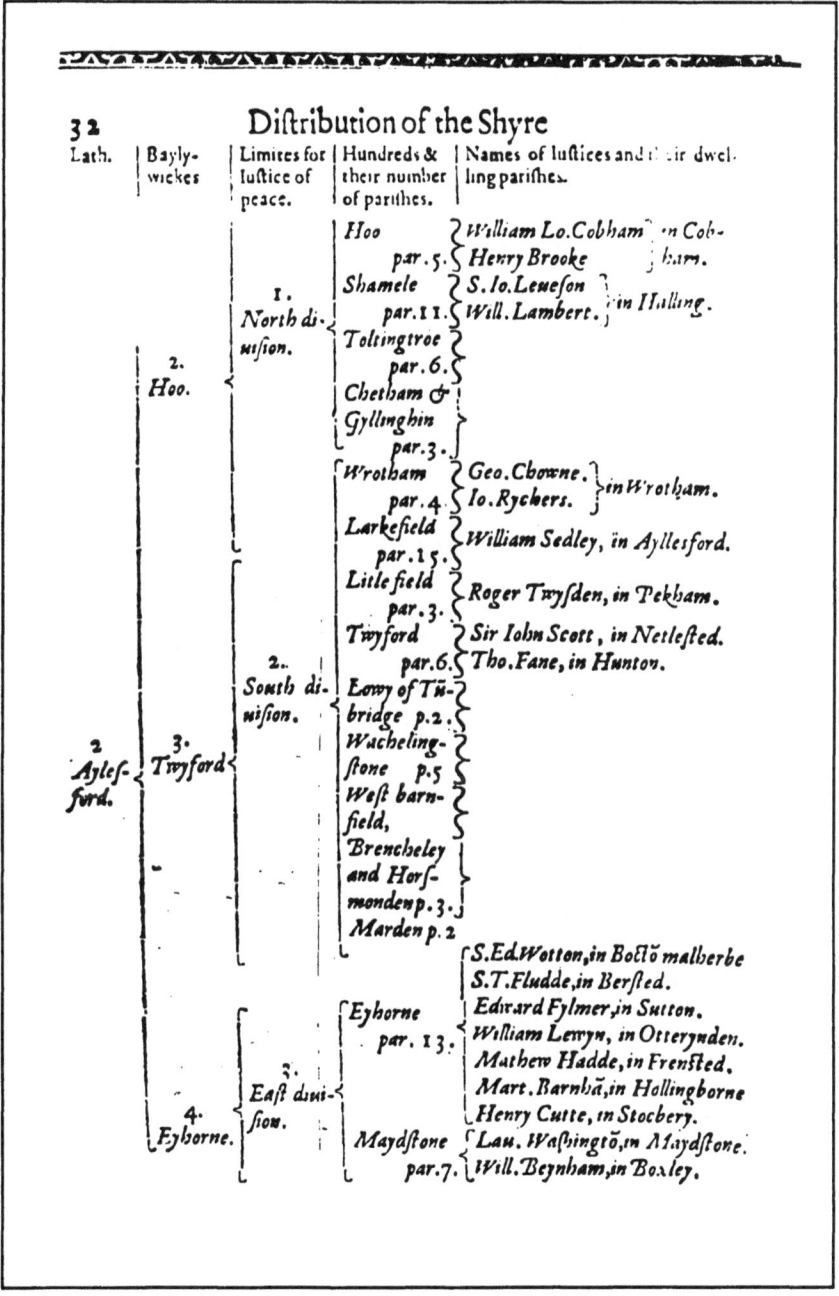

Figure 19. Description of the shire of Kent, displayed as Ramist table, from W. Lombarde, *A Preambulation of Kent* (1598) [31].

for precise communication. In short, while some forms of discourse used Ramist method to improve the organization and format of text, others used bracketed presentation indiscriminately. Throughout the last half of the Renaissance, we can find books that display excesses in both Ramist display and the use of undifferentiated linear discourse.

Theologians, particularly Puritans, were another group who were attracted to Ramist method and often used it injudiciously. For example, the Puritan apologist John Udall, in *A commentarie vpon the lamentations of Jeremy* (1593) [32] incorporated fold-out pages displaying Ramist tables to reveal the major segments of the Book of Jeremiah (see Figure 20). The need for extended pages, usually bound as fold-out pages to capture Ramist dichotomies and partitions, may well have inspired other writers, particularly writers of military how-to books, to use fold-out pages for tables, drawings, and supporting illustrations. This topic will be discussed in Chapter 6, which focuses on graphics and technical description. A noticeable increase in the use of fold-out pages followed the use of such pages for Ramist charts in the later sixteenth century.

James Warre, who wrote both religious and commercial books, used tables rather than discourse to record Scriptural instruction. His religious work, *The Tovch-stone of Trvth, Wherein Veritie, by Scripture is plainely confirmed, and Error confuted* (1620) [33], placed a descriptive heading at the top of each page. Below each heading appeared a table of Bible chapters and verses supporting each truth. Each page was a self-contained table devoted to verses relating to one truth (see Figure 21). Warre used the same tabular organization system in *The Merchants Hand-maide; or, a booke containing tables, for the speedie casting up and true valuing of an commoditie* (1622) [34]. Ramism, with its rigid system of division, showed how truth could be as easily quantified in religion as in commerce. Visualization of truth aided the reader's access, learning, and memory. Ramist form added visual argument to the idea that truth could be grasped, displayed, and contained within the printed page.

After Ramus, technical books, such as navigational books, showed sharply improved methods of spatial display to convey information. Before 1550, most navigation books were composed of undifferentiated text that contained no illustrations. After 1580, however, works showed consistent improvements in page layout—use of white space, drawings, tables, maps, and charts—that provided clear testimony to writers' awareness of the necessity of visual presentation to convey information with precision. A comparison of *The Rutter of the Sea* (1555) [35], Eden's *The Arte of Nauigation* (1579) [36], and Johnson's *The Light of Navigation* (1612) [37] reveals the same substantial improvements in page design that were evident in medical works.

In short, we should not underestimate the influence of Ramus on the development of tables, as navigation books published prior to 1550 used no tables (or poorly designed tables) and minimal visual aids. Eden's *The Arte of Nauigation* [36] was one of the first books to have a world map, a fold-out, in which the New

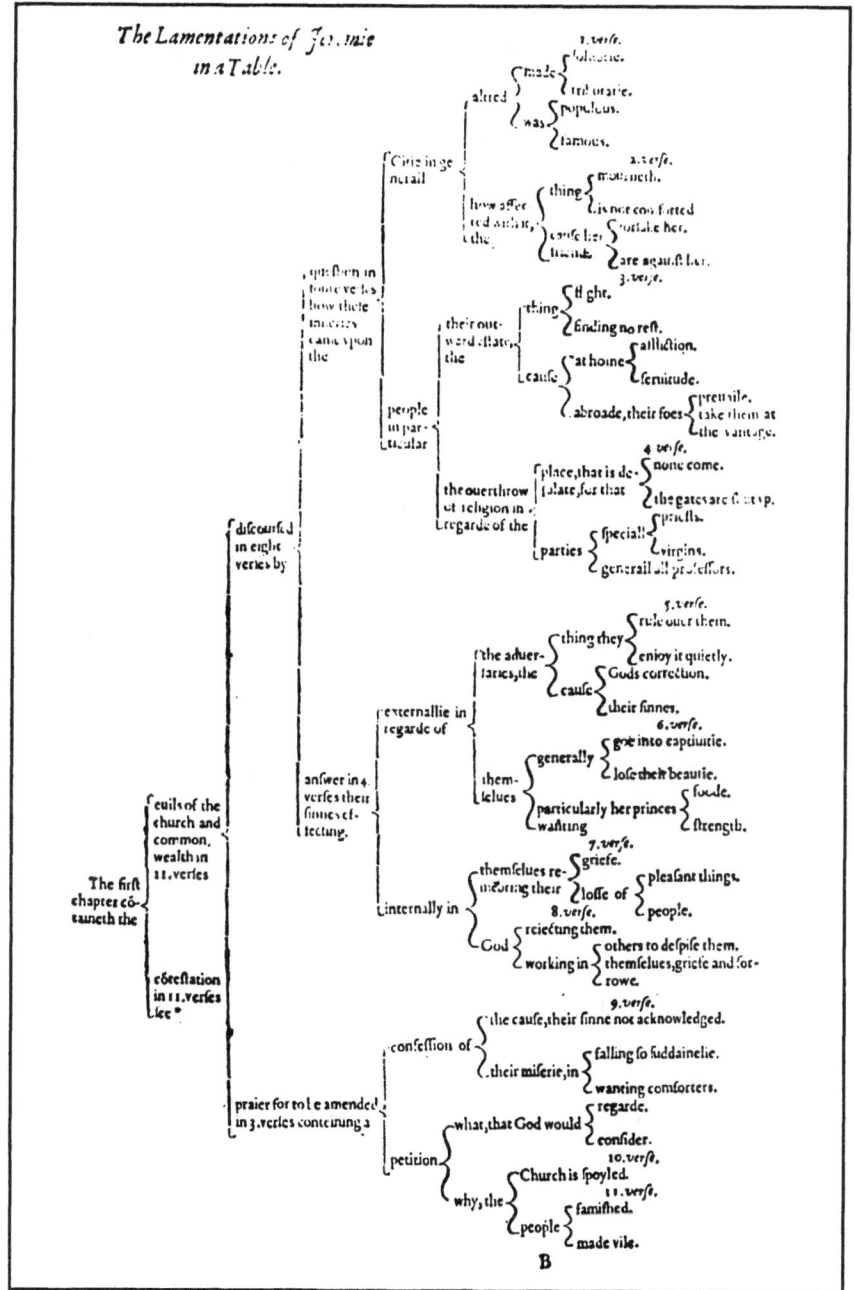

Figure 20. Ramist bracket showing the content of I Jeremiah, from
John Udall, *A commentarie vpon the lamentations of Jeremy* (1599 edition) [32].

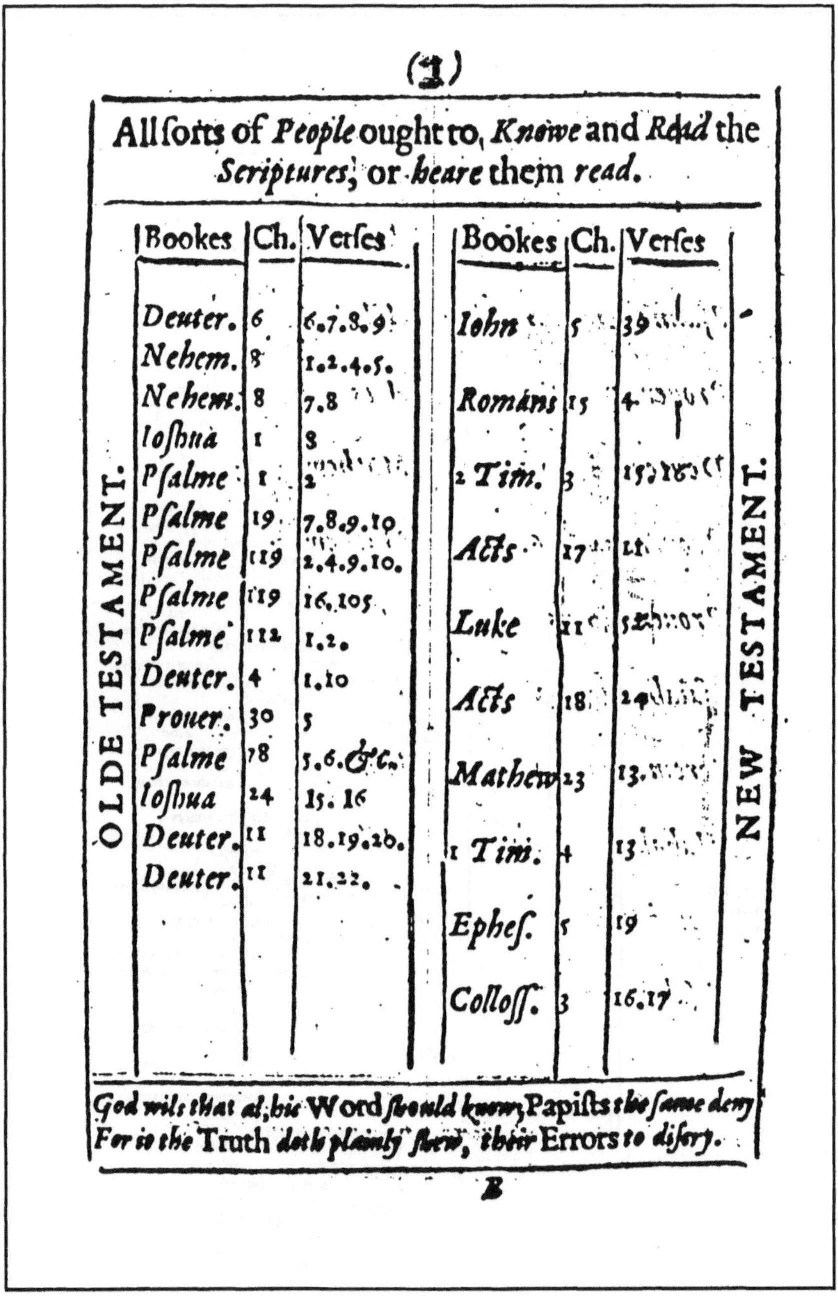

(1)

All sorts of *People* ought to, *Knowe* and *Read* the *Scriptures*, or *heare* them *read*.

OLDE TESTAMENT.

Bookes	Ch.	Verses
Deuter.	6	6.7.8.9
Nehem.	8	1.2.4.5.
Nehem.	8	7.8
Iosbua	1	8
Psalme	1	2
Psalme	19	7.8.9.10.
Psalme	119	2.4.9.10.
Psalme	119	16.105
Psalme	112	1.2.
Deuter.	4	1.10
Prouer.	30	5
Psalme	78	5.6.&c.
Iosbua	24	15.16
Deuter.	11	18.19.20.
Deuter.	11	21.22.

NEW TESTAMENT.

Bookes	Ch.	Verses
Iohn	5	39
Romans	15	4
2 *Tim.*	3	15.16.17
Acts	17	11
Luke	11	28
Acts	18	24
Mathew	23	13
1 *Tim.*	4	13
Ephes.	5	19
Colloss.	3	16.17

God wils that al his Word *should know,* Papists *the same deny*
For to the Truth *doth plainly shew, their* Errors *to discry.*

B

Figure 21. Table of scriptural lists, from James Warre, *The Tovch-stone of Trvth* . . . (1621 edition) [33].

World was delineated on the same sheet with the western parts of Europe and Africa (see Figure 22). This work also includes extensive drawings, astronomical diagrams, and tables giving declinations of the sun. Johnson's *The Light of Navigation* [37] used sophisticated page design techniques and high-quality type (see Figure 23). In this work, patterns for making navigational instruments were printed on loose pages and inserted at the appropriate point in the text where instructions for making the instrument were given. This arrangement allowed readers to remove the pattern page and use it to make the instrument (Figure 23). The deliberate design used in both of these works becomes evident when we compare them with books printed before Ramus. Both works stand in stark contrast to *The Rutter of the Sea* (1555) [35], which, like Moulton's *This is the Myrour or Glasse of helth* [9] (Figure 8), used minimal formatting to demarcate chapter divisions and no visual aids.

Returning briefly to medical works, we see that rigid partition of content was merged with page design techniques. In Alexander Read's *A description of the body of man* (1634) [38], each page presents a titled subclassification of the human body—such as bones of the hands, feet, legs, or skull, or the bones of the jaw. Parts were labeled for easy visual reference (see Chapter 6, Figure 10, p. 196). Because verbal and visual information appeared side by side, the reader could move easily between drawing and written discourse explaining the drawing. Although division and partition had been used in Aristotelian rhetoric, it was Ramus who demanded that division be visually presented to enforce the importance of partition in thought.

How the concept of visualized partition infused medical writing can be seen in *The Frenche Chirurgerye* of Jacques Guillemeau (1597) [39], who used detailed drawings of surgical instruments. Similar to Read's work, *The Frenche Chirurgerye* placed verbal descriptions of instruments across the page from illustrations of the instruments themselves, whose parts were lettered to correspond strictly to the verbal description (see Figure 24). The larger organization of the book also reflects Ramist method with its use of enumerated overviews (advanced organizers) (see Figure 25), similar to those used by Edwards in his work on fevers [27]. While Caxton used introductions, as shown in Figure 1, the demarcation of text into specific segments was enhanced visually and structurally as a result of Ramist insistence on organized partition of content.

Other Design By-Products of Ramus's Visual Rhetoric

Shifts in typography and page design with an emphasis on readability can be seen 1) if we examine other kinds of technical books that enjoyed multiple editions for over a half century and 2) if we examine changes in those works across editions. The first extant published English cookbook, published by Pynson in 1500 [40], when compared with early seventeenth-century cookbooks, makes a similar point. This book, only sixteen pages long, has no title page, minimal visual

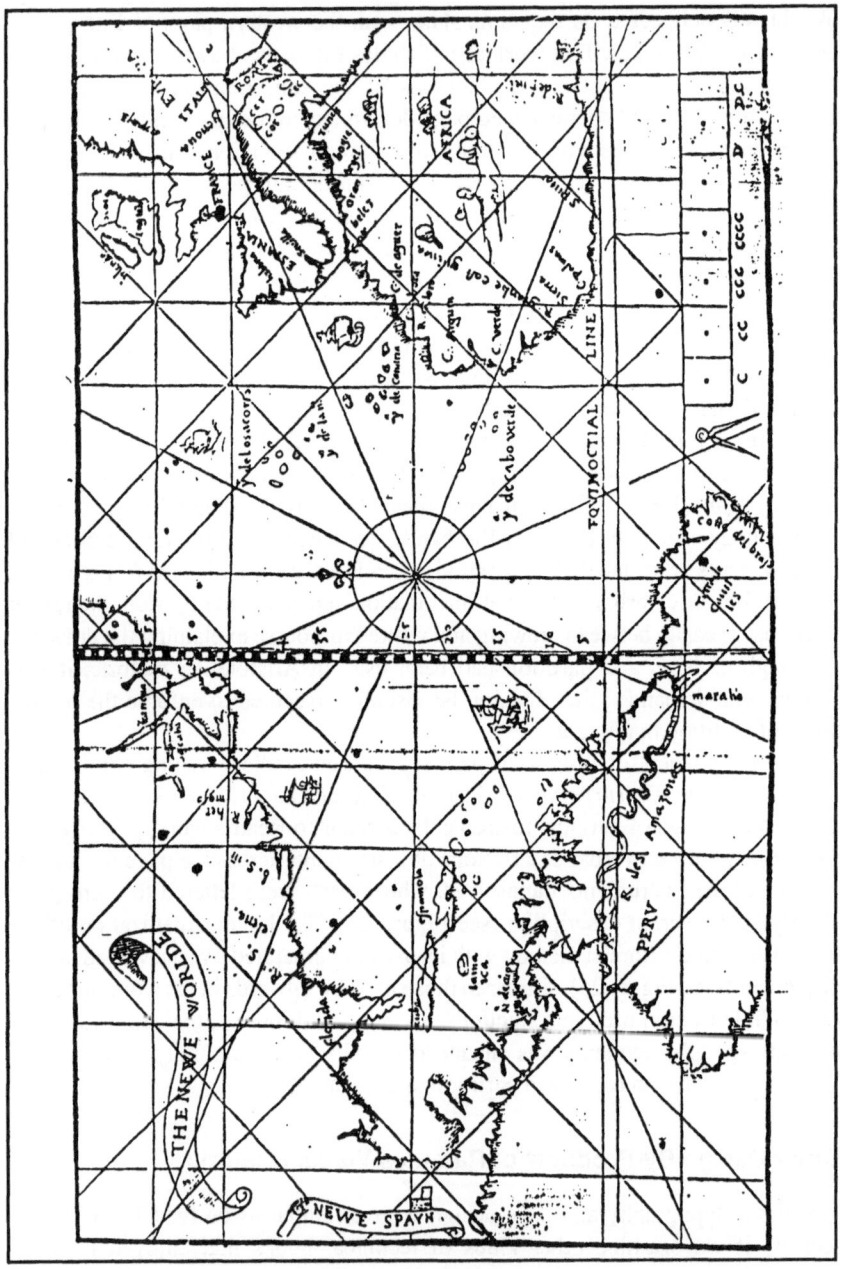

Figure 22. First important published navigational map, from
Richard Eden, *The Arte of Nauigation* (1561 edition) [36].

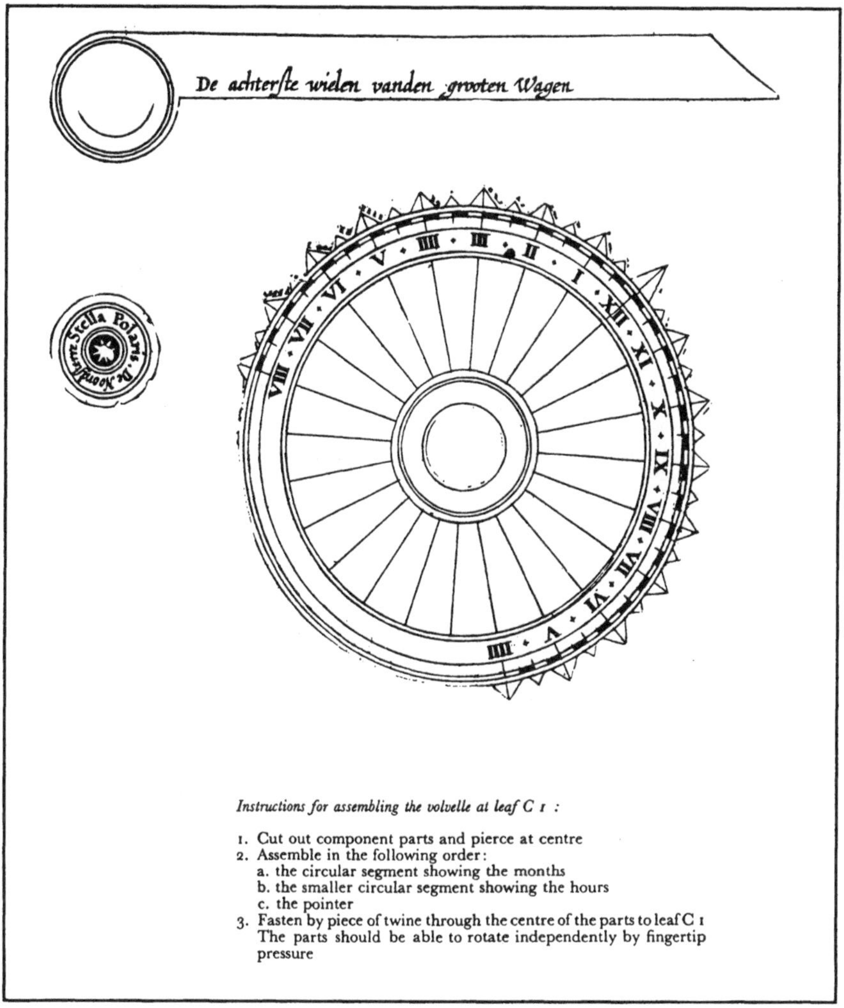

De achterſte wielen vanden grooten Wagen

Stella Polaris

Instructions for assembling the volvelle at leaf C 1 :

1. Cut out component parts and pierce at centre
2. Assemble in the following order:
   a. the circular segment showing the months
   b. the smaller circular segment showing the hours
   c. the pointer
3. Fasten by piece of twine through the centre of the parts to leaf C 1
   The parts should be able to rotate independently by fingertip pressure

Figure 23. Pattern to be used for making a volvelle, from
W. Johnson, *The Light of Navigation* (1612) [37].

access cues, and an aggregative collection of two dozen recipes. In contrast, Platt's *Delightes for Ladies, to adorne their Persons, Tables, closets and distillatories* (1603) [41] used an index (located at the beginning of the book), illuminated borders on every page, and roman type. Similarly, Murrel's *A Daily Exercise for Ladies and Gentlewomen* (1617) [42] used a table of contents which classifies 113 recipes into "pastes, marmalades, conserues, tartuffes, gellies, breads, rough rock candies, sucket candies, cordiall waters, conceits in

The Theſaurarye, or ſtorehouſe

Explicatione of the Characters con-
taynede in the table of theſe Inſtrumētes which
are propre and neceſſarye to the extirpatione of
membres.

A,A,Indicateth the Knife vvhervvith on the ſuddayn,
vve may cutt the ſkinne , and the muſcles to the
bone vvhen vve deſire to extirpate anye mortifyede
Ioynㄑe, or membre. And is of the Latiniſtes callede
Culter exciſorius lunatus, in Engliſhe a ſemiluna-
te cuttinge Knife: it is in this forme compoſed becauſe it might the eaſyer comprehende all the fleſhe in the circuite,

・, Demonſtrateth the ſuperioure parte of the backe,
vvhich after a ſorte ought to be ſomvvhate acute,
& ſharpe, to ſcrape thervvith the Perioſtium from
the bone,vvhich at the firſte time coulde not all at
once be cutte throughe,

B, The perforatione, or hole vvhich is beneath in the
blade,vvhich yealdeth backvvardes into the handle and ther occulteth it ſelfe ther by to contayne
the knife ſteadye.

C , A hole vvhich is in the handle, vvherin is a little
iron barre, vvhich paſſeth cleane throughe vvher-
vvithe the blade is faſtenede.

D The end of the foreſayed blade,through the vvhich
he is impedited to ſtirre backvvardes in effectuatin-
ge of the opration. Some ther are vvhich conten-
te themſelves vvith a common razer,vvhich behinde they involve vvith linnen, leaſte that it ſhoulde
revolute backvvardes , and are of opinione that it
vvould beter be done vvith a razor havinge an emi-
nent bellye,then vvith a ſemilunare knife : and for
confirmatione heerof they take example of the La-
niators,or Bouchers,and of Coockes,vvhich much
rather take a knife vvith an eminente belly , and in
manner,and forme of a razor, then of ſuch an one
vvhich repreſenteth a halfe moone.

E, E, Demonſtrate the Savve vvhich is vvholye a-
mountede, vvith the, Bovve,Blade,and handle, and
is in Latine callede Serra,ſhe is not heere placede in
her magnitude, becauſe the place, can not heere be
ſoe greate,vvhich muſt notvvithſtanding be a good
foote,and tvvo inches of length in her blade, & the
handle foure,or five inches longe.

F,A little pegge of Iron,vvhich houldeth together the
tvvo peeces of the Bovve.

G,An other iron ſerue,vvhich combineth the blade,&
the bovve together.

H,H,Tvvo branches of the Bovve.

I, A viſe.

K,The ſeparatede Handle.

L,The Blade alone ſeparatede.

M, The end of the Bovve , vvhich is clefte, in the
vvhich,the end of the blade vvith the hole therof is
put.

・, The hole,or perforatione,vvhich is in the blade.

, The ſcrue,vvhich muſt have his penetratione cleane throughe the bovve, and the blade, as playnlye
vve may behoulde the ſame notede vvith G,

N,Signifieth the viſe vvhich is occultede in the end of
the handle,vvhich attayneth to the end of the blade
notede vvith O,by that meanes to dravve in the ſame,and faſten the ſayede blade.

O,The end of the blade, vvher there is a ſcrue to receave the viſe.

P,A ſplitte,vvhich is in the blade, to receave therin a
ſcrue notede vvith the figure 4.

4,The ſcrue, vvhich is thruſt cleane throughe the bovve therbye to houlde faſt the blade.

Q, The end of the ſayde Bovve , vvhich is receavede

of the end of the handle , in the vvhich is a ſplitte,
throughe the vvhich the blade paſſeth, vvhen vve
deſire to prepare the ſavves.

1,2,3, Demonſtrate certayne little ſcrues ,to vſe the ſame in time of neceſſitye.

R,A dentifiede , or toothede Crovves bill to clenche
the vaynes,and take houlde theron,the Ioynㄑe beinge extirpatede,and vve deſire to religate the ſayede vaynes,it is in Latine callede Roſtrum Corvi-
num.

S, A reſorte,or ſpringe becauſe it might allvvayes be
aperte.

T, The bille of the ſame,vvhich on his end is rovvnd,
and toothede becauſe the threde might vvith the
more facillitye glide therover vvith out beinge in
anye place ſtayede.

V, The Needle , vvhervvith vve convenientlye may
ſtitch,vvhen vve deſire to religate a Vayne and is in
Latine callede Acus.

X,A hollovve knife L,Culter fiſtularis, G,Syringoto-
me this Inſtrument occludeth it ſelfe in the pipe.

「1, The puncture, or poynㄑe of the ſame, vnder the
vvhich a little bullet of vvaxe is faſtenede , or ſome
other plaſter , leaſt that the poynㄑe ſhoulde hurte
ſome other places vvhen as vve intrude the ſame
in anye fiſtle,thruſtinge as it vveare in anye ſounde
fiſthe,vayne,arterye,or ſinnue,vvith this Inſtrumē-
te in a ſhorte time vve maye deſcide through a greate qvantitye of fleſhe, there are ſome alſo vvhich
vvith this knife at one time cā cut of a greate quantitye of fleſhe.

Y, Reſcindente,inſtrumentes to cut of fingers: L,For-
ceps exciſoria,this inſtrumente muſt be greate, and
ſtronge.

The Finger vvhich is ſpoylede ,and corruptede.

How we ought reſtraygne the bloode
after the extirpatione of a Ioynㄑe, without v-
ſinge anye hot Iron , onlye throughe ligature,
which is of two ſortes, ether with the Crowes-
bille,or withthe Needle.

a,a,Signifye a hippe from the vvhich the legge is ex-
tirpated.

b,b,The Vaynes & Arteryes by the vvhich the bloode
ex ſulteth,and ſpringeth out.

c,cBoth the endes of the Crovvesbille,vvhervvith the
Vaynes are compræhendede, by that meanes to tye
them.

d, The Crovveſbille.

e,e, Both the focilles of the Legge.

f, The ſpringe,or reſorte of the Crovves bille.

g,g, The Arme the fiſte vvherof is extirpatede.

h, The orifices , or mouth of the Vaynes, out of the
vvhich iſſueth bloode.

i, The ſituatione of the Vayne.

l, The place vvher the firſte ſtitch muſt he placede on
the one ſyde of the Vayne: and heere is to be notede,that vve muſt firſt thruſte , in the ſkinne of the
Arme, vvithout vvholye dravvinge throughe the
threde.

m,The evente of the ſeconde ſtitche,vvhich muſt,be
begūne one the other ſyde of the Vayne internally
in the Arme,and muſt pearce throughe the ſkinne.

n, The little compreſſione, vvhich muſt be interſitua
tede betvveene the ſtitches,as tovvardes the lettre a
& the knittinge,theron both the endes of the threde reaſonable ſtiffelye:this little compreſſe, preven-
teth the cuttinge throughe of the threde,and cauſeth noe payne.

Figure 24. Pages of surgical instrument description, from
Jacques Guillemeau, *The Frenche Chirurgerye* (1597) [39].

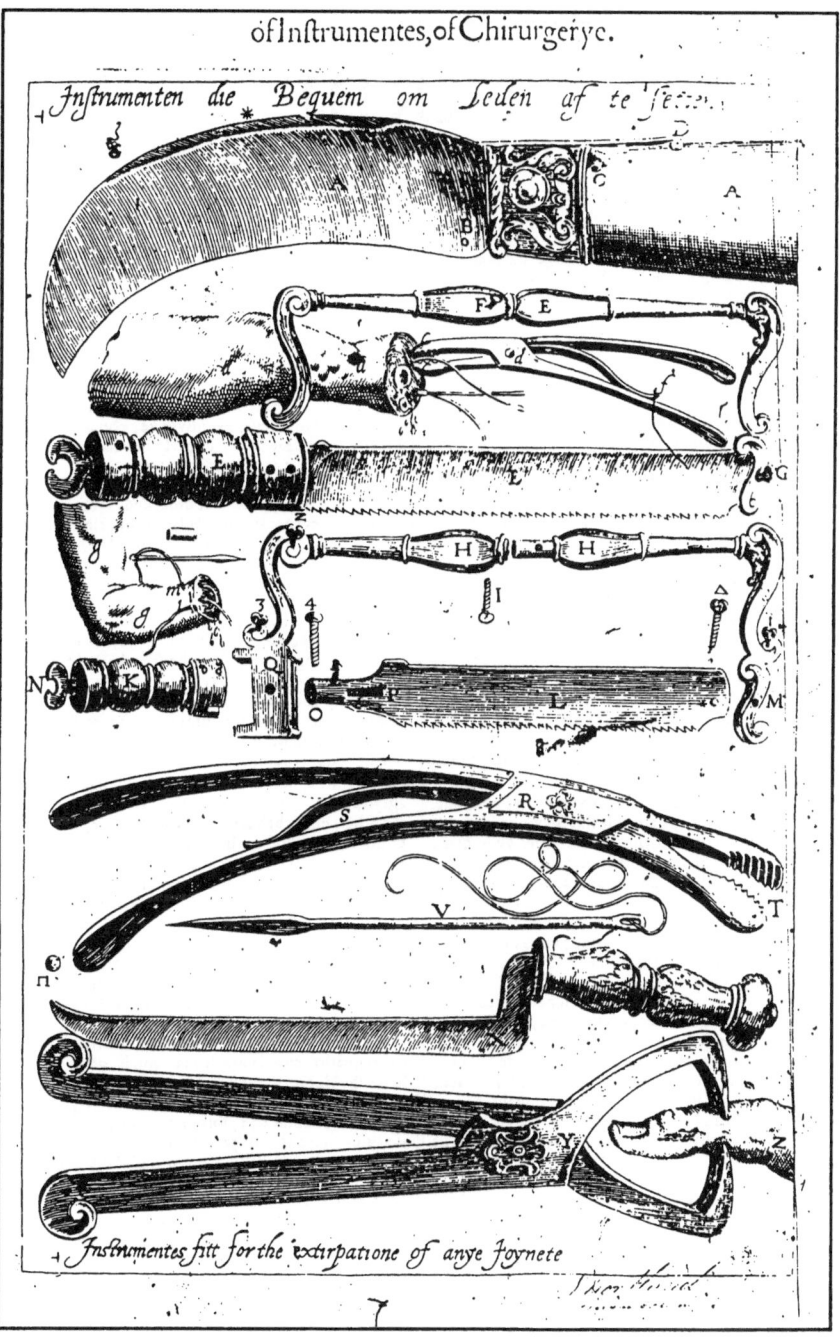

Figure 24. (Cont'd.)

The Frenche Chirurgerye

THE THIRDE TRE-
ATISE OF THE OPERATI-
on of Chyrurgerye, wherin is difcourfede and handelede of
the fovvinge or future of vvoundes. Contayninge fixe Chapters.

VVhat the future, or fovvinge together of avvounde is and the vfe therof.   Chap.1.
VVherone vve muft note in the fovvinge of a vvounde.   Chap.2.
VVhat is needfulle, to the fovvinge, and of the meanes, hovv to doe the fame.   Chap.3.
Of the fpecies, or differences of fovvinge, and of the time to remove the fame.   Chap.4.
Hovv vve ought to repofe agayne the guttes, vvith the net, vvhenas they hange out of the body.   Chap.5.
Of the *Gaftroraphia*, or fovvinge of the bellye-   Chap.6.

What fovvinge is, and the vfe therof, and in what impedimentes fhe is neceffarye, and in what partes. Chap.1.

Sixe thinges vvhich in fovvi nge mutt be conlidered

THe Chyrurgiane ought to confider fixe efpecialle thinges, vvhich concerne the fovvinge of vvoundes: Firft the vfe therof, that is, in vvhat impedimēts it is neceffarye, & in vvhat partes: vvhat vve muft therin confider: vvhat ther is vvantinge to effecte the forefayed fovvinge: after vvhat manner vve ought to doe it, & hovve manye fafhions, & differences ther are of the fame. Therfore fovvinge of a vvounde is *Defcriptione of fovvinge.* nothinge els, but a vnitinge, and couplinge together of the diffevered partes: vvhich vve are contrarye to nature, feparated and parted one frō the other, vvhich fore fayed vnitinge muft be effected vvithe a thredded needle.

Inventione & vfe of fovvinge.
The occafione, vvhy yve in anye vvounde, or feparated parte, vfe this fovvinge, is to vniteagayne, & ioyne them together, vvherof the convenienteft meanes is, this fovvinge, & the vfe of this combinatione, and that efpeciallye in all fuch partes, vvhich vve can not conveniently or aptlye binde together, even as vve may playnlye fee, in the greate vvoundes of the armes, & legges, vvhich are overthvvartlye vvounded, in like manner alfoe in the bodye vvvich happen in the length therof, for the lippes or edges of the fame, are foe much feparated one from the other and caufeth the vvoūde foe to gape, that fhee coulde not vvithout great daunger be cured, vnleffe that in anye place vve fovved it, to bring them together and vnition: becaufe that all incarnate, or flefhye partes of the body, are dravvne throughe vvith certayne finnuifhe fibers or filamētes, the vvhich beinge overthvvartlye, or contradictorylye feparated the one, from the other, the one lippe of the vvoūde is dravven opyvarde, & the other dovvnevvard or the one on the right fyde, & the other one the

left, all accordinge as the vvoūde or the feparatione is more or leffe, overthvvarte, cōtradictorye, lōge, deepe, or vndeepe. In like forte is the fovvinge, verye neceffarye in a vvoūde, vvherin a peece or parcell of flefhe hangeth one the one fyde, and one the other end as yet connexed, even as it commonlye chaunceth moft commonlye, in a great hevve or flafhe, throughe the vvhich the care, hangeth by the heade, or in any other parte, as in the nofe, the vvhich helde faft but at one end onlye.

Sovvinge in a feparated place, is vnprofitable and needeles.
It happeneth alfoe fome times, that the vvhole parte is cleane cut of, vvherin the fovv inge is nothinge profitable, vvherfore therin vve muft not vfe anye fovvinge at all, to cure thē agayne, for then in the feparated parte is noe more life, & therfore can noe more be nourifhed of the bodye, by the vvhich the curinge ought to come.

And althoughe that all thofe partes, vvhich agaynft the courfe of nature are feparated frō the other, might behove to be healed agayne, cā not beare or fuffer to be vnited or coalited, even as are the fynnues, Tendones, & Cartilages or griffells, becaufe after the opinione of aunciente profeffors, and as alfoe the vfe and practife teacheth vs, ther muft confequentlye follovve of one pricke, or thruft vvith a nedle, in the fynnues, or tendones, great payne of all fortes, inflāmatione, convulfione of fynnues, and fome times alfoe death it felfe, throughe the fympathye, or compaffione, vvhich they have vvith there firfte originalle, the vvhich *Galenus*, hath fhevved vnto vs, in a vvoūde, belovve the Hockes in the vvhich, confideringe *The tēdo nes are daungerous to fovve and vvhy.* the greatnes therof, it vvas verye neede full that there fhoulde be a profounde future, & that not onlye to bringe the fuperfituated places, together, but alfoe the profoūde lyinge partes of the vvounde: the vvhich he endevourige to effecte, hath feparated the tēdones frō the mufcles: for becaufe ther is great daūger confifteth in the hurtinge of the fynnues, even in like forte is ther daunger in hurtingo of

sugar-workes, to drie fruits, and physicall receipts." Despite their length, however, cookbooks continued to be published as small, pocket-size quartos or duodecimos that could be easily carried about.

Ramus's influence is distinctly evident in Buttes's *Diets Dry Dinner* (1599) [43], which describes menus, foods, and their dietary advantages and disadvantages. This little book used brackets on the title page to group and list the eight courses presented in the book (see Figure 26A). Each food, as illustrated in Figure 26B, was partitioned in a way that reflects Ramist division: choice, use, hurt, correction, preparation, degree, season, age, and constitution.

Other examples of late Renaissance page design that anticipate modern technical book design can be found in Gervase Markham's *Cheape and Good Hvsbandry* [44], published in 1614. Markham's book begins with a traditional table of contents that uses well-spaced italic and roman type. The table of contents is followed by a glossary, an alphabetized listing of "hard words"—presented in italics with their definitions in roman type. In the main section of the book, long chapters are visually subdivided by exdented headings which announce the key points of the chapter. The book features wide margins, and its organization and format make it look and feel much like a modern technical manual (see Figure 27).

Another excellent example of modern page design can be seen in *The Carpenters Rvle* [45], published in 1602, written for craftsmen who possessed a basic knowledge of arithmetic. The book, divided into three parts, uses ornate chapter divisions, clearly demarcated chapter numbers and titles, and marginal glosses. Chapter numbers and titles are in roman type, but the text itself appears in a high-quality script. Instructions or descriptions of geometric concepts are numbered. Visuals are integrated with the text. Word division occurs at syllables, and the right margin of the text is justified (see Figure 28).

That seventeenth-century printers were increasingly aware of the need for effective visual design is evident in a wide variety of how-to books. For example, Walter Gedde's *A Booke of Svndry Dravghtes* [46], published in 1615, opens with instructions for drawing nine different designs. The verbal instruction appears directly across the page from the design (see Figure 29). The second part of this book is comprised of 103 completed designs, many of which are referenced in the instructions for the designs in the first part of the book. By reading a draught instruction and then turning to the design, the reader can see how a specific design can be fully developed. The last design, number 103, is a large fold-out that shows how four windows using various renditions of designs must look when placed together. This work shows how verbal and visual presentations were designed to complement each other. The final section of the book provides instructions for erecting a glass furnace, annealing the glass, and "receipets" for creating colors. Headings and instructions are well-spaced and visually accessible.

By the closing years of the English Renaissance, 1630-1640, even cheaply printed books showed the influence of sophisticated page design to enhance the

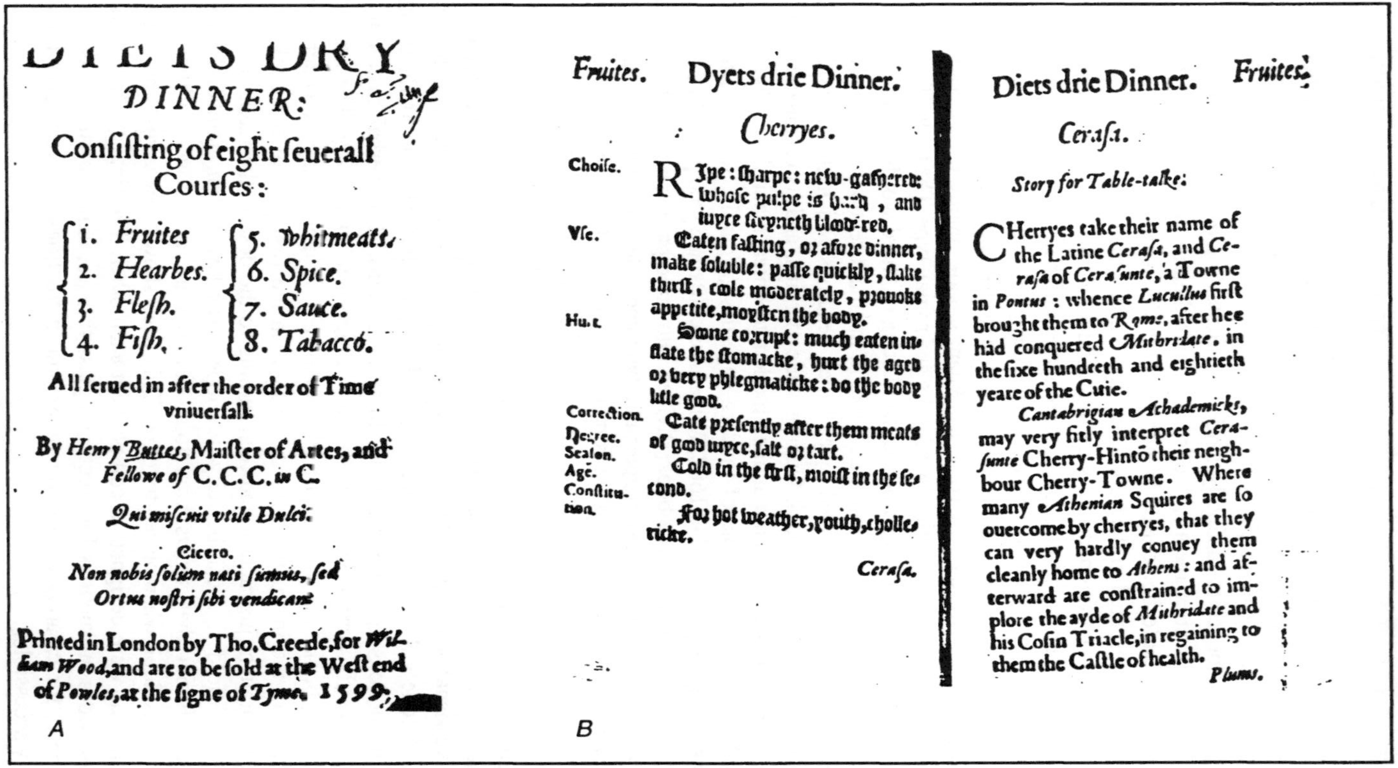

DIETS DRY DINNER:

Consisting of eight seuerall Courses:

1. Fruites
2. Hearbes.
3. Flesh.
4. Fish.
5. Whitmeats.
6. Spice.
7. Sauce.
8. Tabacco.

All serued in after the order of Time vniuersall.

By Henry Buttes, Maister of Artes, and Fellowe of C. C. C. in C.

Qui miscuit vtile Dulci.

Cicero.
Non nobis solùm nati sumus, sed
Ortus nostri sibi vendicant.

Printed in London by Tho. Creede, for William Wood, and are to be sold at the West end of Powles, at the signe of Tyme. 1599.

A

Fruites. Dyets drie Dinner.

Cherryes.

Choise. Ripe: sharpe: new-gathered: whose pulpe is hard, and iuyce sireyneth blood-red.

Vse. Eaten fasting, or afore dinner, make soluble: passe quickly, slake thirst, coole moderately, prouoke appetite, moysten the body.

Hurt. Some corrupt: much eaten inflate the stomacke, hurt the aged or very phlegmaticke: do the body litle good.

Correction. Eate presently after them meats of good iuyce, salt or tart.

Degree. Season. Age. Constitution. Cold in the first, moist in the second.

For hot weather, youth, chollericke.

Cerasa.

Diets drie Dinner. Fruites.

Cerasa.

Story for Table-talke:

CHerryes take their name of the Latine Cerasa, and Cerasa of Cerasunte, a Towne in Pontus: whence Lucullus first brought them to Rome, after hee had conquered Mithridate, in the sixe hundreth and eightieth yeare of the Citie.

Cantabrigian Achademicks, may very fitly interpret Cerasunte Cherry-Hinto their neighbour Cherry-Towne. Where many Athenian Squires are so ouercome by cherryes, that they can very hardly conuey them cleanly home to Athens: and afterward are constrained to implore the ayde of Mithridate and his Cosin Triacle, in regaining to them the Castle of health.

Plums.

B

Figure 26. Ramist influence on organization of a culinary book. *A*, Title page; *B*, example of content page. From H. Buttes, *Diets Dry Dinner* (1599) [43].

Chap. XXII.
Of Bots and Wormes of all forts.

THe Bots and gnawing of VVormes is a grieuous
paine, and the fignes to know them is the horfes
oft beating his belly , and tumbling and wallowing
on the ground with much defire to lye on his backe.
The cure is: take either the feeds bruifed,or the leaues
chopt of the hearbe *Amaea*, and mixe it with Hony,
and making two or three balls thereof,make the horfe
fwallow them downe.

Chap. XXIII.
Of paine in the Kidnyes; paine-piffe, or the Stone.

ALL thefe difeafes fpring from one ground,which
is onely grauell and hard matter gathered together in the Kidnyes, and fo ftopping the conduits of
Vrine : the fignes are onely that the horfe will oft
ftraine to piffe but cannot. The cure is, to take a
handfull of *Mayden-haire*, and fteepe it all night in a
quart of ftrong Ale, and giue it the horfe to drinke
euery Morning till he be well, this will breake any
ftone whatfoeuer in a horfe.

Chap. XXIIII.
Of the Strangullion.

THis is a foreneffe in the horfes yard, and a hot
burning fmarting when he piffeth : the fignes are,
hee will piffe oft, yet but a drop or two at once.
The cure is, to boyle in the water which he drinketh
good ftore of the hearbe called *Mayth* or *Hogs-fenell*,
and it will cure him.

D 3   Chap.

Figure 27. Page of a farming instruction manual, from
Gervase Markham, *Cheape and Good Hvsbandry* (1614) [44].

is the center of the figure. Thus then I deuine the two long prick lines to finde the center of this figure A.

1. The prisme or piece of timber whereof the figure A, or any other irregular mani-sided figure is the base or end, is measured by multiplying the content of the base by the length of the piece, and the product giues the solid content.

2. Or els, take the base of all the sides or compasse about as for one side of a squared piece of timber, and the said perpendicular for the other side, and so measure it by any of the waies taught in the second part.

CHAP. 9.

How irregular mani-sided figures are measured.

Irregular mani-sided figures, as also regular mani-sided figures, may be thus measured. Deuide the figure into triangles, as here I haue diuided the figure B by certen prick lines. Then measure each triangle by it selfe, as is taught in the third Chapter of this part, and adde the contents of all the Triangles together; and their summe or totall is the content of the figure.

1. The prisme or piece of timber, whose base or end is the figure B, or any other irregular mani-sided figure, is measured by multiplying the content of the base or end in the length of the piece, and the product giues the solid content.

2. Or els, take the content of the base, and so measure it by any of the waies taught in the second part of this booke.

3. And because for the measuring of euery sort of timber

I haue vsed to set downe some way by which it may be done without Arithmeticke: Therefore for a third way, as euery triangle in the figure B is measured, so measure so many followers toward of the triangles be the bases, and whose lengths are y length of the irregular piece of timber, as is taught in the third Chapter of this third part; and then adde all their contents for the content of the whole piece.

CHAP. 10.

What a Circle, a Semicircle, and a Sector are, and how they be measured.

1. Circle is a figure, plaine and round.
2. The round line y makes the Circle is called the Circumference.

3. The point in the middle of the circle is called y Center.

4. The line (and indeed euery line) passing by the Center to both sides of the Circumference, is called the Diameter, as y black line in this circle C.

5. Halfe the Diameter, and indeed euery line passing from the Center to the Circumference, is called the Semidiameter.

6. All lines in the Circle (except the Diameter) which are drawne from side to side, are called Cords. Such is the prickt line in the circle C.

7. The Diameter cuts the circle into two equall parts; each of which parts is called a Semicircle.

8. A cord cuts the circle into two vnequall parts; each of which parts is called a Section of a Circle. The one is called the greater section, because it is greater then y Semicircle.

The

Figure 28. Pages showing effective instructional format, from R. More, *The Carpenters Rvle* (1602) [45].

Directions how to make your Square.

First then concerning the square, which although it be cōmon to sundry artificers, each one hauing his owne forme and deuise, in drawing of it, some by deuision of halfe circle, some by other rules: there is herefore set downe a more perfect and readier way for any worke in draught, and also if neede be, to try the square rule thereby: for certainely, glasse worke of all others, requires most an exact square, for the distance, of one haire out of square, will deface the whole worke, and bring it out of all frame.

The ordering of the square.

To drawe this square: First you must drawe your line draught so long as your worke requireth, marking right theron with your compasse, three points of alike distance, next, deuide your compasse larger, setting the one point on the figure 2. & with the other drawe a quarter circle betwixt 4. and 5. after, remoue your compasse to the figure 3. and drawe a crosse circle ouer betwixt 6. and 7. and wheresoeuer the iust middle of the crosse appeareth, marke the same with the point of the compasse, as you may perceiue in the character, by the figure 8. lastly, drawe a draught, betwixt the figure 1. and the crosse point by the figure 8. and thou shalt find a perfect square to conduct thee all thy draughts. As on the former side is shewed.

A 3. The

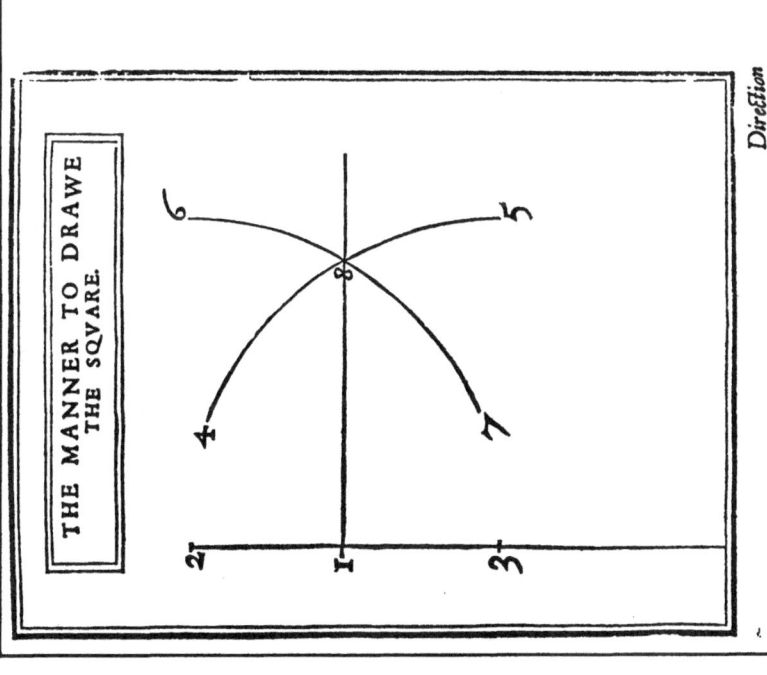

THE MANNER TO DRAWE
THE SQVARE.

Figure 29. Instructions for "glass draughts," from Walter Gedde, *A Booke of Svndry Dravghtes* (1615) [46].

83

visual accessibility of the contents. For example, James Achesone's *The Military Garden* [47] (1629) is a brief handbook (36 pages) for military drill. The book opens with descriptions of the responsibilities of each rank and visual depictions of how companies should march. This little book is particularly interesting because it uses typography and visual cues to explain how oral commands are to be given. For example, the actual commands appear in italics, with the explanation in roman type (see Figure 30).

ECHOES OF RAMIST METHOD IN MODERN TECHNICAL WRITING PHILOSOPHY AND PRACTICE

Today much research of interest to technical writers focuses on ways in which format and page design can reveal organization and content to enhance the readability of documents. As the examples in this chapter have shown, many current practices in document design reflect changes that surfaced from the influence of Ramist logic as well as improvements in typography during the second half of the English Renaissance.

First, research conducted since 1970 has found that readers read selectively, often for reference only. As a result, technical writers and document designers now recognize the power that format and page design give the writer in controlling how readers see and process organization and content. Modern readability research, echoing Ramist method, recognizes that readers, in order to process ideas effectively, must be able to *see* the text—infer content from titles, headings, and the organization of material. Modern document design researchers such as Huckin [48] also echo Ramus's view that revealing content aids the reader in searching the text—searching through it or skimming it—to determine content before beginning to read evaluatively or analytically. Slater, citing studies by Meyer, Brandt, and Bluth and Taylor, states that "readers who can identify and used expository text structure and/or main ideas remember more of what they read than do readers who cannot or do not" [49, p. 198].

Second, current readability research also echoes Ramus's belief in the importance of establishing a priority for ideas. Research has established that readers read hierarchically, that main points should be partitioned and placed in a more dominant position than subordinate points. As Huckin [48], Slater [49], and Meyer and Rice [50]—to name only a few of a large number of document design researchers—have all noted, research now tells us that "the important points of a text should be placed in superior positions hierarchically: in headings, in subheadings. . . . If certain deails are also important, they can be listed instead of subordinated" [48, p. 95]; and, "Main ideas are more memorable than supporting ones" [49, p. 199]. Ramus and his English disciples would be pleased that research has shown that hierarchical structure, including listing, enhances recognition of main ideas (the levels effect). Thus, the use of titles, headings, and subheadings—which grew with the increased emphasis on organizing text to reveal content quickly—

22.	*The Militarie Garden.*

or vpon any suddaine assault to cause face to any hand, or charge to any hand.

The company standing in battell orderly, both in ranks and strings, and true distance in both, the words of exercising are these, viz.

Faces to the right.	*To your first order.*
Faces to the left.	*To your first order.*
Faces right about to the reare.	*To your first order.*
Faces left about to the reare.	*To your first order.*

PRESENTING.

Present to the right.	*To your first order.*
Present to the left.	*To your first order.*
Present right about to the reare.	*To your first order.*
Present left about to the reare.	*To your first order.*
Faces to the right and left by division.	*To your first order.*
Faces to the front and reare by division.	*To your first order.*

PRESENTING.

Present to the right and left by division.	*To your first order.*
Present to the front and reare by division.	*To your first order.*

There is diverse sorts of exercising of Musquetiers by themselues apart, which cannot well bee set downe but in action.

The first is to winne ground vpon the enemie.

The second is to lose ground, or making a retreate, and

yes

Figure 30. Military commands, from James Achesone, *The Military Garden* (1629) [47].

has become the hallmark of modern technical writing. Ramus would be equally pleased that modern readability research has shown that informative headings improve reading comprehension [51], especially headings that reveal organization and direct the reader in accessing the text [49].

Third, schema theory also echoes Ramus. Just as Ramus saw his method as a means of linking new situations to ancient knowledge, of advancing the best of Aristotle, schema theory researchers such as Anderson and Pearson [52] and Meyer [53] recognize the importance of using textual devices to activate and sustain the reader's prior knowledge to bring it to bear on the processing of new information. Thus, we now use advanced organizers, many designed just as they were in Ramist texts, to prepare readers for the information that will follow. To Ramus's disciples we owe the summary, usually diagrammatically charted, appearing across the page from the prose narrative that developed the content according to clearly partitioned segments (Figures 17 and 25). Perhaps one of the greatest contributions to readability by Ramus's disciples was their use of textual effects to make text accessible and appropriate to the audience, another important finding in modern readability research such as that by Faigley and Witte [54], Duin and Penn [55], and Duin [56].

By the late sixteenth century, a fourth important characteristic of page design was also surfacing with increased regularity: effective use of white space, use of tables to visually quantify information, increased use of detailed drawings—all of which have been shown to enhance the readability of documents [51, 57]. Thus, by the end of the English Renaissance, we can see the emergence of the highly visual how-to text which anticipated the central place that sophisticated graphics now command. Ultimately, however, to Ramus we owe a major step in the emergence of textuality, to the text that would stand alone, a legal requirement of many forms of modern technical writing. Like a box, the document must contain all necessary information for the reader to perform the process safely; it must be clear apart from any vocal intercession. Ong's assessment of Ramus's method [14] also applies to modern technical writing. In both Ramist rhetoric and modern technical writing, written communication is visual perception from which auditory perception has been removed. With the independence of text, the role of person-to-person communication became unnecessary. The visual display of knowledge became monologic rather than dialogic, with the goal of text as single meaning emanating from the text to the mind of the reader.

Howell credited Roger Ascham, in his 1550 letter to Sturm, with being the first Englishman to recognize Ramus [13]. Ironically, however, this letter also marked Ascham as one of the earliest English educators to discuss the importance of clarity and readability in writing, a quality that Ascham apparently recognized in Ramus's attacks on scholastic educational methods. In this letter, it was Aristotle's lack of readability on which he commented specifically. Ascham's criticism of Aristotle could apply to modern criticism of bad technical writing:

The excellent doctrine of Aristotle seems too devoid of adornment, too obscure, for delight in reading it to be able to arouse the zeal of the many, or for the usefulness of it to compensate for the labors involved, because almost everywhere it is taught without the accurate use of examples [quoted in 13, p. 174].

Current research in visual design, emphasis on the visual transmission of data and facts, the growing emphasis on graphics—tables, flowcharts, circle graphs, bar graphs, and creative graphics—all attest to our continuing belief in the persuasiveness of the visualized text, which will be further discussed in Chapter 6. This belief in extreme textuality, as Ong has clearly shown [4, 14], resulted in part from advances in print technology; but Ramism, with its emphasis on texts as teaching instruments, made an important contribution to the need for texts that provided clear truth to practitioners as well as student readers [14, p. 309]. Research suggests that we owe much of that difference to Ramus.

REFERENCES

1. C. DePisan, *The Fayt of Armes & of Chyualrye*, Westminster, 1489 [STC 7269].
2. D. J. Berners, *The Boke of Saint Albans*, Saint Albans, 1486 [STC 3308].
3. R. Hand, *English Hawking and Hunting in the Boke of St. Albans*, Oxford University Press, Oxford, 1975.
4. W. J. Ong, *Orality and Literacy: The Technologizing of the Word*, Methuen, London and New York, 1982.
5. L. Kellner, *Historical Outlines of English Syntax*, Macmillan, London and New York, 1892.
6. H. Von Braunschweig, *The Vertuouse Boke of Distyllacyon of the Waters of All Maner of Herbes*, London, 1527 [STC 13435].
7. H. Von Braunschweig, *The Noble Experyence of the Vertuous Handy Warke of Surgeri*, London, 1525 [STC 13434].
8. T. Paynell, *A Moche Profitable Treatise Against the Pestilence*, London, 1534 [STC 24226].
9. T. Moulton, *This is the Myrour or Glasse of helth*, London, 1539 [STC 18214].
10. D. Smyth, *Hereafter foloweth the Judgemet of all Urynes*, London, 1540 [STC 14834].
11. A. Borde, *The Breuiary of Helthe*, London, 1547. Reprinted in *The English Experience*, Number 362, Theatrum Orbis Terrarum, Amsterdam and New York, 1971.
12. W. Bulleins, *Bulleins Bulwarke of Defece against all Sicknes, Sornes, and Woundes, that Dooe Daily Affault Mankinde*, London, 1562 [STC 4033].
13. W. S. Howell, *Logic and Rhetoric in England, 1500-1700*, Russell & Russell, New York, 1961.
14. W. J. Ong, *Ramus, Method, and the Decay of Dialogue*, Harvard University Press, Cambridge, Massachusetts, 1958.
15. P. Miller, *The New England Mind: The Seventeenth Century*, Harvard University Press, Cambridge, Massachusetts, 1954.
16. P. Ramus, *P. Rami Dialecticae Libridvo*, Cantabrigiae, 1584 [STC 15243].
17. T. Elyot, *The Castle of Helth*, London, 1536-1539 [STC 7643].

18. T. Elyot, *The Boke Name the Gouernour*, London, 1531 [STC 7635].
19. J. Banister, *A Needefull, new, and necessarie treatise of Chyrurgerie, briefly comprehending the generall and particular curation of Vlcers*, London, 1575 [STC 1360].
20. W. Clowes, *A Profitable and Necessarie Booke of Obseruations, for all those that are burned with the flame of Gun power, &c.*, London, 1595 [STC 5442].
21. T. Gale, *Certaine workes of chirurgerie*, Four Parts, London, 1563 [STC 11529].
22. J. Guillemeau, *A Worthy Treatise of the eyes; contayning the knowledge and the cure of one hundreth and thirteene diseases, incidens vnto them*, London, 1587 [STC 12499].
23. M. R. Makkylmenaeum, *The Logicke of the Most Excellent Philosopher P. Ramus Martyr, Newly translated, and in diuers places corrected, after the mynde of the Author*, London, 1574 [STC 15426].
24. R. Caldwell (trans.), *Tables of Svrgerie*, by H. Morus, London, 1585 [STC 18204].
25. A. Read, *The chirurgicall lectures of tumors and ulcers delivered in the Chirurgeons Hall*, London, 1635 [STC 20781].
26. E. Edwards, *The analysis of Chyrurgery, Being the Theorique and Practique part therof*, London, 1636 [STC 7511].
27. E. Edwards, *The Cvre of All Sorts of Fevers*, London, 1638 [STC 1712].
28. J. E. Murdoch, *Album of Science: Antiquity and the Middle Ages*, Scribner, New York, 1984.
29. E. Vaughan, *Ten Introductions: How to read, and in reading, how to vnderstand; and in vnderstanding, how to beare in mind all the bookes, chapters, and verses, conteined in the holie Bible*, London, 1594 [STC 24599].
30. V. Leigh, *The Moste Profitable And commendable Science, of Surveiyng of Landes, Tenementes, and Hereditamentes*, London, 1578 [STC 15417].
31. W. Lombarde, *A Preambulation of Kent: Conteining the Description, Hystorie, and Customes of that Shyre*, London, 1596 [STC 15176].
32. J. Udall, *A commentarie vpon the lamentations of Jeremy*, London, 1593 [STC 24494].
33. J. Warre, *The Tovch-stone of Trvth. Wherein Veritie, by Scripture is plainely confirmed, and Error confuted*, London, 1620 [STC 25090].
34. J. Warre, *The Merchants Mand-maide; or, a booke containing tables, for the speedie casting up and true valuing of an commoditie*, London, 1622 [STC 24908].
35. P. Garcia, *The Rutter of the Sea*, London, 1555 [STC 11552].
36. R. Eden, *The Arte of Nauigation*, London, 1579 [STC 5800].
37. W. Johnson, *The Light of Navigation*, London, 1612 [STC 3110].
38. A. Read, *A Description of the Body of Man*, London, 1634 [STC 20783].
39. J. Guillemeau, *The French Chirurgerye*, London, 1597 [STC 12498].
40. [Book of Cookery], R. Pynson, London, 1500 [STC 3297].
41. H. Platt, *Delightes for Ladies, to adorne their Persons, Tables, closets and distillatories*, London, 1603 [STC 19978a].
42. J. Murrel, *A Daily Exercise for Ladies and Gentlewomen*, London, 1617 [STC 18301].
43. H. Buttes, *Diets Dry Dinner: Consisting of eight seuerall Courses*, London, 1599 [STC 4207].
44. G. Markham, *Cheape and Good Hvsbandry*, London, 1614 [STC 17336].
45. R. More, *The Carpenters Rvle*, London, 1602 [STC 18075].
46. W. Gedde, *A Booke of Svndry Dravghtes*, London, 1615 [STC 11695].

47. J. Achesone, *The Military Garden, Or Instrvctions For All Young Sovldiers*, London, 1629 [STC 88].

48. T. Huckin, A Cognitive Approach to Readability, in *New Essays in Technical and Scientific Communication: Research, Theory, Practice*, P. Anderson, R. J. Brockmann, and C. R. Miller (eds.), Baywood Publishing, Amityville, New York, pp. 90-110, 1983.

49. W. H. Slater, Current Theory and Research on What Constitutes Readable Expository Text, *Technical Writing Teacher, 15*, pp. 195-206, 1988.

50. B. J. R. Meyer and G. E. Rice, The Structure of the Text, in *Handbook of Reading Research*, P. D. Pearson (ed.), Longman, New York, pp. 315-351, 1984.

51. D. B. Felker, *Document Design: A Review of the Relevant Research*, American Institute of Research, Washington, D.C., 1980.

52. T. C. Anderson and P. D. Pearson, A Scheme-Theoretic View of Basic Processes in Reading Comprehension, in *Handbook of Reading Research*, P. D. Pearson (ed.), Longman, New York, pp. 255-291, 1984.

53. B. J. F. Meyer, Organizational Aspects of Text: Effects on Reading, Comprehension and Applications to the Classroom, in *Promoting Reading Comprehension*, J. Flood (ed.), International Reading Association, Newark, Delaware, pp. 113-138, 1984.

54. L. Faigley and S. P. Witte, Topical Focus in Technical Writing, in *New Essays in Technical and Scientific Communication: Research, Theory, Practice*, P. Anderson, R. J. Brockmann, and C. R. Miller (eds.), Baywood Publishing, Amityville, New York, pp. 59-70, 1983.

55. A. H. Duin and P. Penn, Identifying the Features that Make Expository Text More Comprehensible, *Contemporary Issues in Reading, 2*, pp. 51-57, 1987.

56. A. H. Duin, How People Read: Implications for Writers, *Technical Writing Teacher, 15*, pp. 185-194, 1988.

57. M. M. Sebrechts, J. G. Deck, and J. B. Black, A Diagrammatic Approach to Computer Instruction for the Naive User, *Behavior Research Methods and Instrumentation, 15*, pp. 200-207, 1983.

CHAPTER 4

Renaissance Technical Books and Their Audiences: Writers Respond to Readers

Titles and the use of distinctive page design provide initial evidence that technical writing existed as a distinct genre in the English Renaissance. Deliberate use of format and page design by writers and printers to enhance the *readability* and accessibility of technical information leads to a conclusion discussed in Chapter 3: Many Renaissance writers were aware of the context in which readers would access a work, whether it would be read for reference (as in the case of medical or military instructions) or carefully and meditatively (as in the case of religious works). Given this evidence that Renaissance technical writers adapted page design to reading context, a further question may be asked: Do these technical books provide added evidence that Renaissance technical writers and printers conscientiously used other methods to make their works appealing? This chapter will show that English Renaissance technical writers were aware of their readers' literacy levels, education, and information needs and developed their texts with these needs in mind.

WRITERS, READERS, AND THE PROFIT MOTIVE

The commercial motive that lay behind publication provides a useful measure of the kind of books that printers believed would sell. As Bennett noted, even Caxton responded to the need for texts upon a range of practical matters: "To this end the printers spread their net wide, using every means in their power to find out what their readers wanted and how they could most successfully be attracted" [1, p. 54]. As Louis Wright also suggested, "The publishers of Elizabethan

England could no more live by the custom of learned and aristocratic readers alone than can modern followers of their trade" [2, pp. 92-93]. Given that printers and writers sought to ensure a steady flow of profitable works, how did they make these works attractive to their intended readers? A study and a comparison of different types of technical books suggest at least three answers.

First, the early printers made available a wide range of books for a growing, increasingly diverse English reading public. But within specific genres of works, particularly technical books, we can track differences among the targeted audiences of these books by the methods that writers used to present their subject matter to their readers.

Second, an increasing number of books on all subjects became available *in English* through translation from popular continental European books or original manuscripts designed for the needs of English readers. While pride in the vernacular was clearly a driving force in the generation of non-Latin texts, so was the growth of a literate English public eager for works in English. The increasing number of English books published throughout the Renaissance suggests that literacy among English readers was growing. The success of these books—in terms of demand—can in some sense be judged by the number of extant editions recorded by the *Short-Title Catalogue*. However, as Hirsch reminds us, many more editions of these works have probably been lost because they were heavily used and were cheaply printed and bound to ensure affordability [3, p. 11]. Such an assessment can be made because of the worn pages of extant copies.

Third, printers provided technical books that, from a modern perspective, fall into four categories: 1) those written to appeal to both a general audience and an expert audience, 2) those for a general audience, 3) those written for two specified audiences (men and women), and 4) those written solely for the expert reader. This wide range of books for a variety of audiences provided the best means of ensuring profit. As Bennett noted, printers saw that the market for information books for expert audiences was limited, and the greatest profit was to be had in providing books that would appeal to a wide readership [1, p. 109; 4, p. 183]. Nonetheless, the *Short-Title Catalogue* records technical books exemplifying all four categories. An analysis of presentation methods of the four types of books provides useful background for tracking and highlighting what have come to be known as standard methods for adapting material for readers of the work. At the same time, these books for readers with varying knowledge levels tell us a great deal about the literacy of men and women readers, as well as the growth of knowledge in many fields, throughout the Renaissance.

TECHNICAL BOOKS FOR
GENERAL AND EXPERT AUDIENCES

The Renaissance technical writer, as many prefaces and title pages suggest, saw the value of addressing a broad spectrum of readers possessing a range of reading

abilities. A common technique was to combine both Latin text or vocabulary with an English translation. This Latin-English combination would thus appeal to readers having grammar school backgrounds as well as readers with only minimal English reading skills acquired in petty school or by home schooling.

Works on Land Management

This Latin-English technique was successfully used by John Fitzherbert, whose *Here begynneth a ryght frutefull mater: and hath to name the boke of surueyeng and improumetes* enjoyed twelve editions between 1523 and 1577 [5]. In the preface, Fitzherbert states the purpose of the work: "to thentent that the lordes, the freholders nor their heyres, shulde not be disheryt, nor have their landes loste nor imbeseled, nor encroached by one from another." This book, a small octavo or quarto (depending on the edition) which could be easily carried about, was advocated by printer Thomas Berthlet in the 1539 edition as useful "for all states that be lordes and possessioners of landes, and for the holders or tenaunes of the same landes to have dayly in hande, to knowe and beare away the contentes." While many landowners would have studied Latin in grammar school and at a university, many landowners from the commercial classes would have been proficient in English only. Fitzherbert apparently saw that his work would be needed by both types of readers.

Fitzherbert opened each chapter with a lengthy descriptive title. The discussion begins with a Latin legal concept, then explains the term in English for landowners who could not read Latin. Each chapter explains specifically how a manor should be surveyed to avoid boundary disputes. The work also contains models for preparing documents and advice on procedure in manorial courts. Because it could be read and used by a wide range of landowners, the work lost little of its value during the fifty-four years it was reprinted.

The opening segment of Chapter 5 of the *Boke of surueyeng*—"Of parkes and bemeyne woodes the whiche the lorde maye assarte and to do his profyte and how many acres they conteyne and what the vesture of an acre is worth and what the grounde is worthe whan the vesture is fallen"—is followed by a Latin rendition of the topic. Fitzherbert then writes in English:

> It is to be enquered/of ykes and demeyne woode/the whiche at the lordes wyll may be affected and plucked up or fallen downe' And how many acres are conteyned in them/and for howe moche the vesture of euery acre may be solde/and how moche the grounde in hym selfe conteyneth when the wode is fallen' and howe moche euery acre is worthe by it selfe by the yere. This is to be understande/of parkes and demyne woode that be in seueraltie/ whereof the lorde at his pleasure/may affect/stocke up by the rootes or falle by the erth/plowe and sowe to his moost profyte as he wyll/And howe many acres of woode are conteyned in the same, for in a

> parke or woode may be two hundred acres and more/and yet nat past a
> hundred acres thereof woode . . . [5].

Notice that this 1523 version uses verticules (virgules) instead of punctuation.
Later editions would use periods to demarcate sentences, but the additive style
(additive sentoids) reminiscent of the oral tradition remained. The style is clearly
that of non-aureate English with its use of farming terminology (demeyne, ykes)
that would have been recognizable to readers who were landowners.

Medical Books for Practitioners with Varying Levels of Education

Because the audience for medical books for specialists was limited, printers and
medical writers learned the commercial value of providing medical books that
would appeal to both the "Latined" specialist and the "un-Latined" English
physician who bought English versions and translations of Italian medical books
to improve his expertise. Physicians who received their training in Italy or France
learned medicine in Latin. Returning to England, they soon saw that communicating medical procedures in the English vernacular was one thing, but finding
English equivalents for a variety of medical terms was quite another. The lack of
English equivalents thus led English physician writers educated on the continent
to adopt Latin and Greek medical terms for use in English medical books. These
writers presented concepts by using the Latin term or phrase followed by the
English equivalent.

These instructional medical books thus raised the knowledge level of
vernacular-speaking physicians but would still appeal to practitioners who had
formal training in continental medicine where Latin and Greek medical terminology was taught and regularly used. As Johnson stated in his discussion of the
controversy over Latin versus English usage in medical books, advocates of
English were initially forced to give way to Latinists [6]. These physicians,
because of their continental medical educations, argued for the superiority of
Greek and Latin for communicating medical knowledge. The view was based on
the fear that medicine's reputation and prestige would be lowered by vernacular
medical books. Latin protected medicine from intrusion by hordes of unlearned
English physicians who had not received proper training. However, un-Latined
native doctors could and did rely on selected works that had been "Englished."

The problem that English medical writers faced in attempting to provide
accurate vernacular medical books—which were mainly composed of technical
descriptions and procedures for physicians and surgeons—is evident when we
examine the anatomical books. Vesalius's work on human anatomy, written
entirely in Latin and first printed in 1542, became the seminal work on anatomy
used in the training of physicians in Italy [7]. When an abridged version of
Compendioso Totius Anatomie Delaneatio Aere Exarata was translated into
English by Nicholas Udall in 1553 [8], the result of Udall's rigidly omitted Latin

terminology was evident: specific anatomical parts, such as the muscles, lost their Latin designations. Without equivalent English terms, anatomical entities were described by clumsy phrasing:

> The fyrst muscle of mouynge the shynne, who doeth degenerate unto that broade tendon, which is bewrapped with the muscles that compasse aboute the thighe, but he is not so thicke but that the muscles vnder him maye ryghte wel be sene.

> The seconde muscle of mouynge the shynne
> The first muscle of mouynge of the shyyne

> The ryght muscle of the second parte of mouinge the head.

> The muscle of stretchynge furth the wreste with a forked tendon [8].

Figure 1 shows a technical description of the human urological system. Note the lack of English medical terms for naming each part. This book was written by one of the most famous textbook writers of the sixteenth century, Robert Record [9], who was fully committed to providing a range of English books. As the English translations of technical descriptions from Vesalius's anatomical drawings reveal, English description without a nomenclature lacks precision. However, the drawings to which these descriptions referred—for example, see Chapter 6, Figures 7 (p. 190), 8 (p. 191), and 9 (pp. 194-195)—at least introduced the un-Latined physician to the major Renaissance research in medicine and anatomy and prepared practitioners to accept the error of Galenic anatomy.

By 1634, prominent English physician writers such as Alexander Read [10] were providing anatomical books which partitioned the human anatomy into subsections and major components of these subsections. Read used either the Latin or the Greek medical term and an English equivalent (see Chapter 6, Figure 10, p. 196), using English descriptive equivalents or the Latin or Greek term in whatever combination was most practicable.

Read's purpose, to help English physician readers gain a working knowledge of essential Latin and Greek terms, can be more clearly seen in his work on tumors, which is well worth careful scrutiny in terms of definition strategy. Figure 2 shows how he combined English, Latin, and Greek to show the parts of surgery. In the following passage Read defines proper Latin terms and warns readers of the differences between Latin and English. He also uses partition, classification, explication, description, and analogy in expanding his definitions making every effort to tie into his readers' knowledge by using the term "choler" as they understood it:

> But before I enter into a particular discourse of an Erysipelas, give mee leave to touch three points briefly, which will make all things which I shall deliver, more plaine. The first, is the signification of this terme choler: The second shall bee of the divers kindes of choler: The third shall set downe the sighnes of a cholerick person: for such an one is most subject to this Tumor.

OF VRINE.

Of the Instrumente and partes by whiche
Vrine is engendred and passeth, marke
[this fygure folowinge.

A. ys the lyuer.
B. the hollowe
vayne.
C. vaynes by
which the reanes
wo drawe the v-
rine, and therfore
be called suckinge
vaynes.
D. the reanes.
E. the water
pypes.
F. is the blad-
dẽr.
G. the spoute of
the parẽ.

All the other
partes besyde, ap-
pertaine to Gene-
ration and seede.

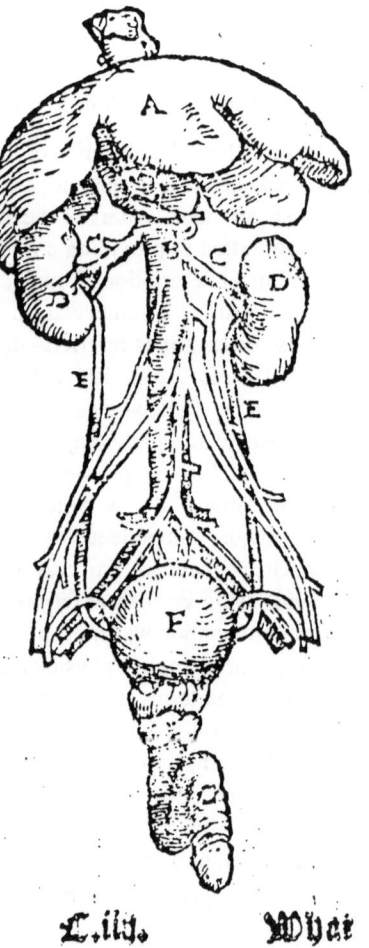

C.iij.   Whẽ

Figure 1. Page showing human urological system, from Robert Record,
The vrinal of physick (1547) [9].

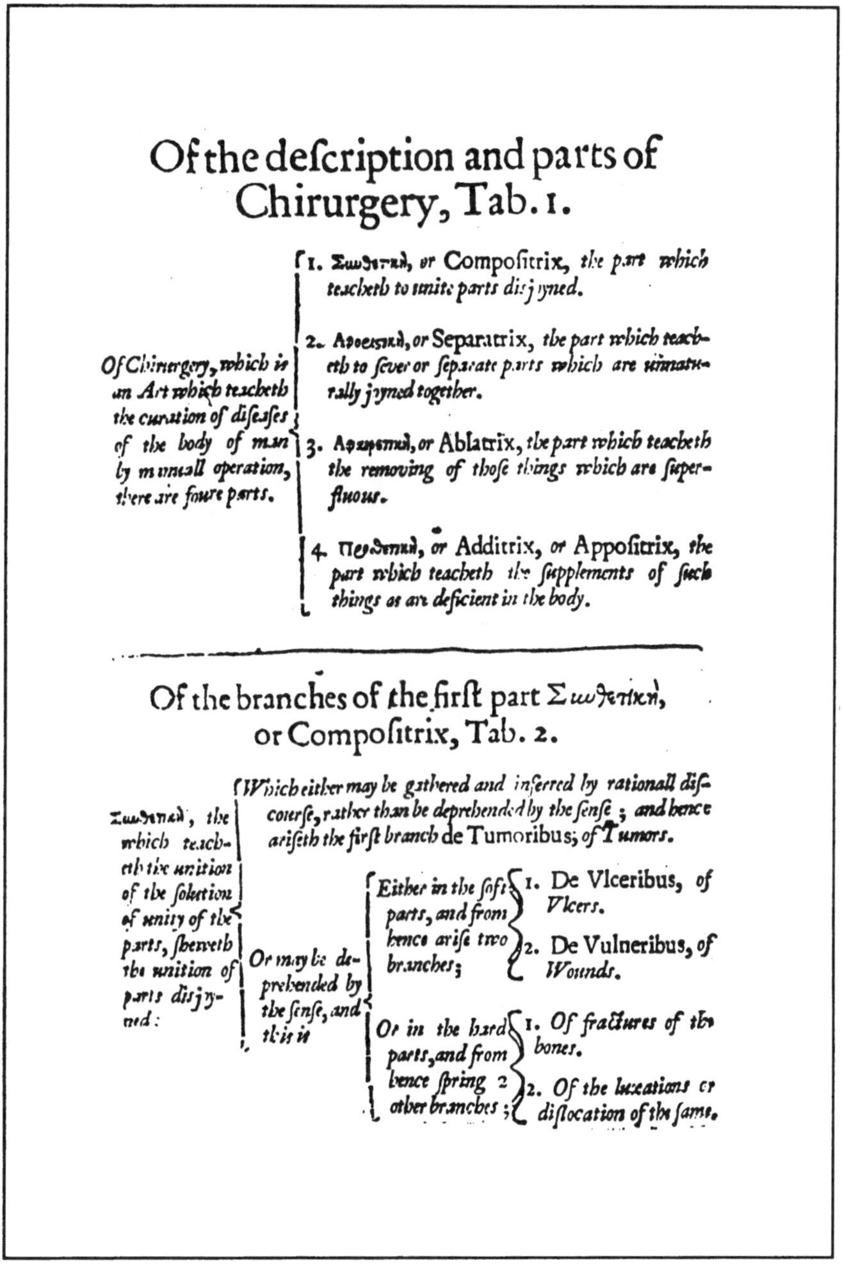

Of the defcription and parts of Chirurgery, Tab. 1.

Of Chirurgery, which is an Art which teacheth the curation of difeafes of the body of man by manuall operation, there are foure parts.

1. Σωδετικά, *or* Compofitrix, *the part which teacheth to unite parts difjoyned.*

2. Αφοεισκά, *or* Separatrix, *the part which teacheth to fever or feparate parts which are unmaturally joyned together.*

3. Αφαιρετικά, *or* Ablatrix, *the part which teacheth the removing of thofe things which are fuperfluous.*

4. Περθετικά, *or* Additrix, *or* Appofitrix, *the part which teacheth the fupplements of fuch things as are deficient in the body.*

Of the branches of the firft part Σωθετικά, or Compofitrix, Tab. 2.

Σωδετικά, the which teacheth the unition of the folution of unity of the parts, fheweth the unition of parts disjoyned:

Which either may be gathered and inferred by rationall difcourfe, rather than be deprehended by the fenfe ; and hence arifeth the firft branch de Tumoribus; of Tumors.

Either in the foft parts, and from hence arife two branches;
1. De Vlceribus, of Vlcers.
2. De Vulneribus, of Wounds.

Or may be deprehended by the fenfe, and this is

Or in the hard parts, and from hence fpring 2 other branches ;
1. Of fractures of the bones.
2. Of the luxations or diflocation of the fame.

Figure 2. Description of a tumor using Latin, Greek, and English, from Alexander Read, *The Chirurgicall Lectures of Tumors and Vlcers* (1635 edition) [11].

Colera then in Latin, and κολερα in Greeke signifieth not the humor, which in English is called choler or the gall; but a disease, whereby this humor is expelled vehemently by vomit and siege. Neverthelesse in our vulgar speech, and with the barbarous late writers, it is taken for the humor it selfe, and so I shall bee inforced to use it. Thus much I thought to insinuate to you, that you should not bee ignorant of the proper and learned significations of the terms of Art.

No *Bilis* or *Fel* in Latine, κολη in Greeke, choler or gall in English is either alimentaraie, or superfluous: the alimentarie is the fourth part of the masse of bloud dry and hot, which more properly might bee called bilious bloud: the superfluous, which serveth not for the nourishing of the body, is either naturall, or unnaturall: the naturall is that, which in the liver is separated from the blood, and turned to *vesicafellis*. The unnaturall comprehendeth foure sorts, under it. I. is the Vitelline, like vnto the yolkes of egges yellow; but thicker and hotter than that of the gall: II. is *Porraces*, which in colour representeth the leeke, it is engendered in the stomack by reason of cruditiee and eating of such meat as engender it as garlick, onions, leeke, and milkin. Infants, who often seege discharge such as humor [11, p. 57].

Read, whose format techniques were examined in Chapter 3, also used a Ramist table to summarize his lecture on "A Treatise of Tumors" [11], as shown in Chapter 3, Figure 17 (p. 66), which also reveals how he attempted to explain tumors in ways that his readers could understand while familiarizing them with proper terminology. Note that in the second paragraph of the prose description for "A Treatise of Tumors" (p. 67) he used Greek and Latin terminology to enforce his English definition of a tumor. His syntax can probably be called utilitarian, what we might describe as "plain style," with its emphasis on subject-verb linear patterns and common speech phrasing (e.g., "Because things bended cause a sticking out") to vivify the Latin and Greek names for his English readers. Read did a masterful job of defining by merging English with Latin descriptive terms.

John Banister wrote a work entitled *A Needefull, new, and necessarie treatise of chyrurgerie, briefly comprehending the generall and particular curation of Vlcers* (1575) [12]. Aside from pharmaceutical instructions, always written in Latin, Banister was careful to avoid medical terminology that would be unfamiliar to his English physician readers. This work clearly illustrates the rudimentary level of ophthalmology. While he used Ramist method, the lack of disease etiology limited his descriptions of disease causation to rudimentary surface description of what could be seen by visual observation. For example, Banister described "Cancer in the Head, Necke, Shouldlers, Brest, Arme-hold, Flanke":

A Cancer non-ulcerate in the brest, is an exceeding great swelling, not yeeld-ing when it is touched, unequall, cruell, like a savage best, piercing & sticking fast within, stretching his rootes very broad, bound as it were with veynes, swelling like *varix* round about: the colour is ashy;, somewhat declinig to purple, of darke blue, soft to the sight, but in feeling most hard, hauing pricking paine every where so largely stretched, that by consent therof it hath

procured pestilent Buboes in the arme-holes, which have reached unto the shoulders [12].

Richard Banister, in *A Treatise of One Hundreth And Thirteen Diseases Of the Eyes, and Eye-Liddes* (1662), described an ulcerate cancer in the following way. While his descriptive method was limited to visual observation, he chose specific, tactile adjectives to aid the reader in identifying the ulcer:

> It is an Vlcer, seruing vnequall, filthy, with swolne, hard, knottie, turned, high, hollow edges, lothsome to behold, abounding with blacke putrified matter, blue, sometime redde and bloodie, from whence floweth continually a thinne, waterish, blacke, yellow, stinking humour. It is named malignant and fierce, of wilde and fierce beasts. For it is a stubborne dissease, made worse by healing and handling [13, p. 443].

Despite both writers' efforts at precise description, the passage (and the entire work) again illustrates the problems that medical writers and translators confronted when they attempted to write only in English and to attain precision when scientific methods of description and causation were as yet unknown.

The extent of Latin usage in medical books often becomes the way of identifying the target audience, if the author's preface does not provide that information. Books for physicians that gave instructions for pharmaceutical preparations used Latin pharmacological symbols and nomenclatures, while medical books for general readers used basic narrative recipes. For example, Clowes's *A Profitable and Necessarie Booke of Obseruations* . . . [14] and John Banister's *A Needefull, new, and necessarie treatise of Chyrurgerie* . . . [13], both written for doctors, provide listed medicinal preparations using Latin quantities and ingredients. In contrast, Thomas Dawson's *The good huswifes Iewell* narrates preparations of folk medicine remedies that would be interesting and readable to a broad non-medical audience [15]. Figure 3 from Banister [13] and Figure 4 from Dawson [15] allow a comparison of these different approaches to medical writing, each clearly designed for the practical information needs and reading comprehension level of its intended audience.

Medicine's dependence on herbal medicine was evident in books providing detailed distillation methods for producing a range of quintessences thought to have healing benefits. Conrad Gesner's *The newe iewell of health* (1576), a 500-page octavo, provides technical descriptions of distillation devices and procedures for making oils and waters from an extensive array of plants, animal parts, and insects [16]. Gesner's work differs from those written for non-medical audiences in its detailed content and preparation procedures that utilized the best of Renaissance distillation equipment and processes. Gesner assumed that physicians would be the only readers with the equipment, expertise, and time necessary to prepare these medical distillations. While many of the ingredients differ little from those found in books for housewives, the text emphasizes the preparation

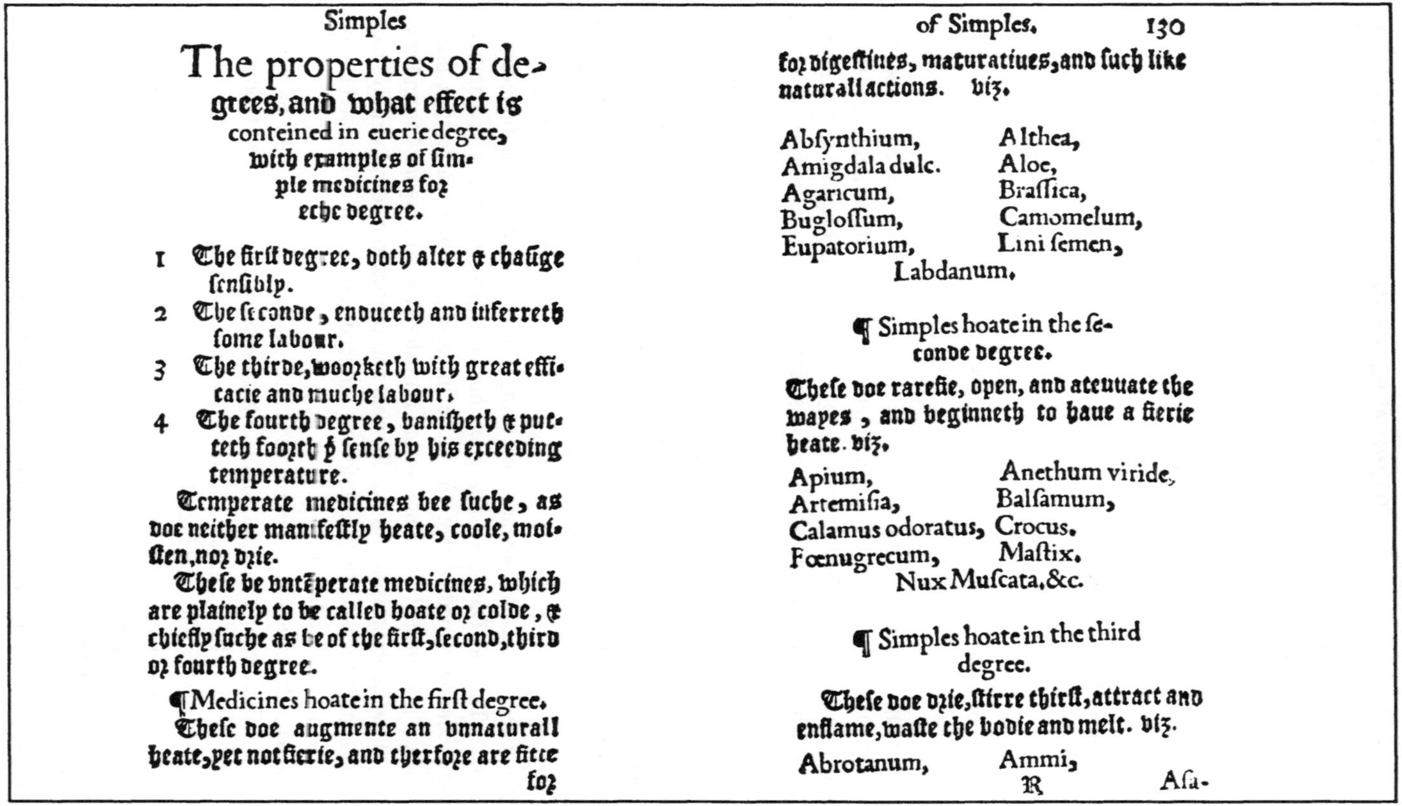

Figure 3. Medical instructions for physicians, from John Banister, *A Needefull, new, and necessarie treatise of Chyrurgerie*... (1575) [13].

Simples

The properties of degrees, and what effect is contained in euerie degree, with examples of simple medicines for eche degree.

1 The first degrees, both alter & change sensibly.
2 The seconde, enduceth and inferreth some labour.
3 The thirde, woorketh with great efficacie and muche labour.
4 The fourth degree, banisheth & putteth foorth ẏ sense by his exceeding temperature.

Temperate medicines bee suche, as doe neither manifestly heate, coole, moisten, nor drie.

These be intemperate medicines, which are plainely to be called hoate or colde, & chiefly suche as be of the firſt, second, third or fourth degree.

¶Medicines hoate in the first degree,
These doe augmente an vnnaturall heate, yet not fierie, and therfore are fitte for

for digestiues, maturatiues, and such like naturall actions. viz.

Absynthium, Althea,
Amigdala dulc. Aloe,
Agaricum, Braſſica,
Bugloſſium, Camomelum,
Eupatorium, Lini ſemen,
Labdanum,

¶ Simples hoate in the seconde degree.

These doe rarefie, open, and ateuuate the mapes, and beginneth to haue a fierie heate. viz.

Apium, Anethum viride,
Artemiſia, Balſamum,
Calamus odoratus, Crocus,
Foenugrecum, Maſtix,
Nux Muſcata, &c.

¶ Simples hoate in the third degree.

These doe drie, stirre thirſt, attract and enflame, waſte the bodie and melt. viz.

Abrotanum, Ammi,
℞ Aſa-

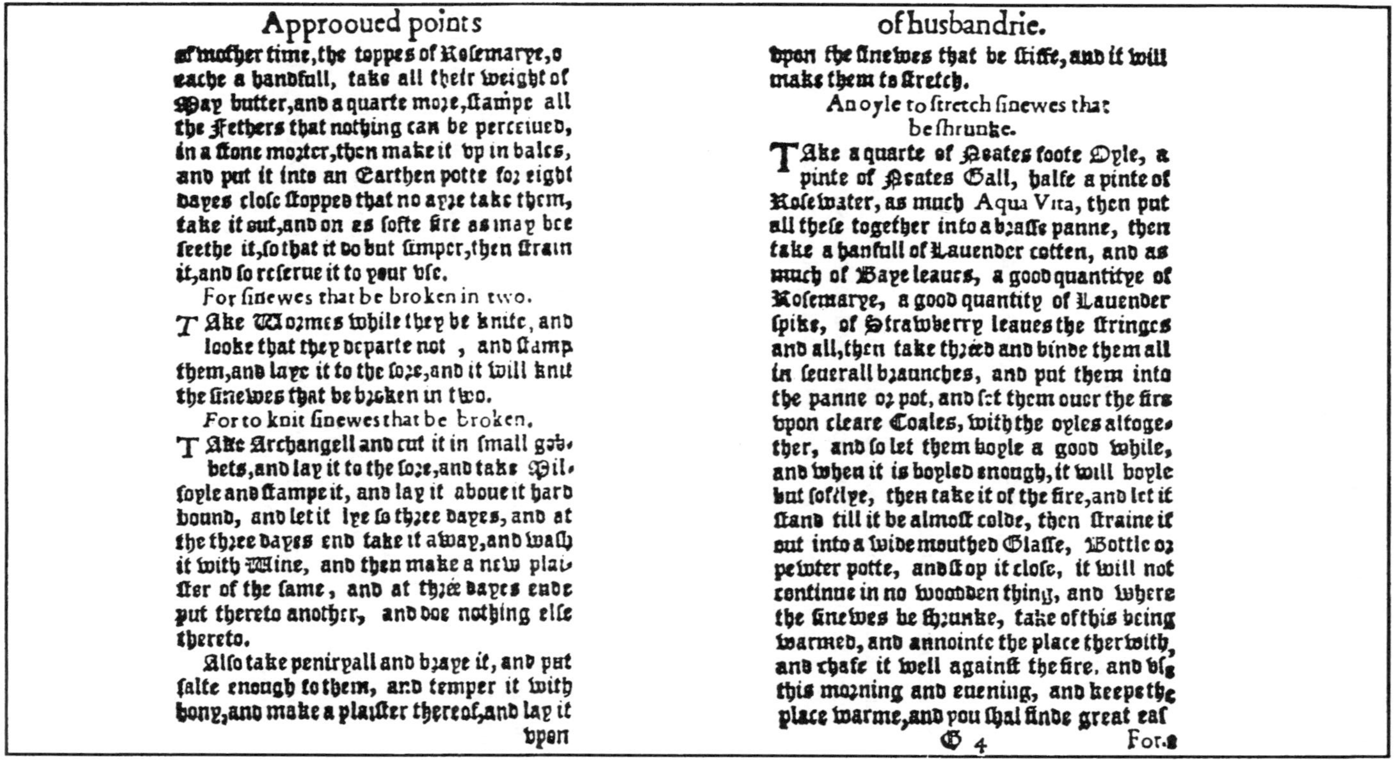

or winter time, the toppes of Rosemary, of eache a handfull, take all their weight of May butter, and a quarte more, stampe all the fethers that nothing can be perceiued, in a stone morter, then make it vp in balls, and put it into an Earthen potte for eight dayes close stopped that no ayre take them, take it out, and on as softe fire as may bee seethe it, so that it do but simper, then straine it, and so reserue it to your vse.

For sinewes that be broken in two.

TAke Wormes while they be knit, and looke that they departe not, and stampe them, and laye it to the sore, and it will knit the sinewes that be broken in two.

For to knit sinewes that be broken.

TAke Archangell and cut it in small gobbets, and lay it to the sore, and take Wilfople and stampe it, and lay it aboue it hard bound, and let it lye so three dayes, and at the three dayes end take it away, and wash it with wine, and then make a new plaister of the same, and at three dayes ende put thereto another, and doe nothing else thereto.

Also take penirall and braye it, and put salte enough to them, and temper it with hony, and make a plaister thereof, and lay it vpon

vpon the sinewes that be stiffe, and it will make them to stretch.

An oyle to stretch sinewes that be shrunke.

TAke a quarte of Neates foote Oyle, a pinte of Neates Gall, halfe a pinte of Rosewater, as much Aqua Vinæ, then put all these together into a brasse panne, then take a handfull of Lauender cotten, and as much of Baye leaues, a good quantitye of Rosemarye, a good quantitie of Lauender spike, of Strawberry leaues the stringes and all, then take thred and binde them all in seuerall braunches, and put them into the panne or pot, and set them ouer the fire vpon cleare Coales, with the oyles altogether, and so let them boyle a good while, and when it is boyled enough, it will boyle but softlye, then take it of the fire, and let it stand till it be almost colde, then straine it out into a wide mouthed Glasse. Bottle or pewter potte, and stop it close, it will not continue in no woodden thing, and where the sinewes be shrunke, take of this being warmed, and annoint the place therwith, and chafe it well against the fire, and vse this morning and euening, and keepe the place warme, and you shal finde great ease

G 4   For &

101

Figure 4. Medicinal instructions for lay readers, from Thomas Dawson, *The good huswifes Iewell* (1596) [15].

process, as illustrated in the distillation apparatus pictured and described in Figure 5. Instructions in these herbals, compared with those written for housewives, are more technical; the products of these processes were more potent (and, one hopes, more beneficial) than homemade preparations. In short, pharmaceutical nomenclature and intricacy of procedure become a significant indicator of intended readership as well as an indicator of the state of knowledge in a particular discipline.

TECHNICAL BOOKS FOR GENERAL AUDIENCES

Self-Help First Aid Manuals

Some Renaissance technical writers crafted their books for specific readers. Distinct differences appear when we compare medical books for physicians with self-help medical books for general readers who sought help in diagnosing their afflictions when a physician was not readily available. Doctors were scarce and financially out of reach for all but the most affluent. Self-help books provided the best source of help aside from orally transmitted folk remedies, many of which were textualized from remedies dictated to the writer. The conversational style of these manuals suggests the oral pattern on which many of these remedies were textualized.

The popularity of basic self-help medicinal books is evident from the large numbers of books of this type that emerged from the London presses from 1580 to 1640, either in translation from popular continental works or in collections of English medicinal recipes. Small, pocket-sized first-aid books comprised a quarter of all editions published and nearly a third of pocket-sized works published between 1486 and 1604 [17, p. 247].

These little books provided descriptions of conditions and a medicinal remedy, usually some sort of food or herbal preparation. As discussed in Chapter 2, *The Castle of Helth* by Sir Thomas Elyot appeared in sixteen editions between 1539 and 1610 [18]. Andrew Borde's *The Breuiary of Helthe* [19] enjoyed eight editions between 1540 and 1582, and *Here begynneth a newe boke of medecynes intytulyd the treasure of pore men* [20] appeared in fifteen editions between 1526 and 1575. Thomas Moulton's *This is the Myrour or Glasse of Helth* was published seventeen times between 1531 and 1580 [21]. *The regiment of life* [22] was published nine times (1544-1596); *The haven of health* [23], seven times (1584-1636); and *A Rich Store-House or treasury for the Diseased* [24], six times (1596-1631). *The Birth of Mankinde . . .* , the only English book on childbirth written for midwives, appeared eighteen times between 1540 and 1614 [25]. Numerous procedural works for analyzing urine to determine sickness also appeared; *Here Beginneth the Seynge of Urynes*, the most popular, appeared in thirteen editions between 1525 and 1562 [26].

of Distillations. 2 6

The Letter A. in this figure repeſenteth the Furnace where the fire appeareth be made and kyndled : the Character B. expreſſeth the Funnell or Chymney of the Furnace: the note C.declareth the Potte ſette and ſtandynge ouer the fire, in whyche the water boylinge is contayned : the Figure D. ſheweth the Pype, by whiche the water boyling runneth forth into a Wooden Tubbe, ſtanding nygh to the Furnace : the letter E. expreſſeth the Tubbe of woode, which recepueth the water heated, wythin which is ſet and ſtandeth the Cucurbite or Bodie of Glaſſe : the letter F.demonſtrateth the Bozia or Cucurbite with his Helmet, which contayneth the matter to be dyſtilled : the figure G.repreſenteth the hollowe Pype, by which the water runneth forth into another waſte Tubbe or Panne ſtanding vnder : the letter H. ſheweth the Glaſſe veſſell, which recepueth the water dyſtilled. It ſæmeth vndoubtedly (ſayth the worthie Geſnerus) the ſame to be the better faſhion of all others, for the Dyſtilling in Balnco Mariæ, but much more commodious, than if the fire were putte vnder the Dyſtilling veſſelles. Conſider and marke the other forme, lyke in a maner to this, hereafter among the Dyles.

¶ The Dyſtillation of the Quinteſſence, in
Balneo Mariæ.

The.xij. Chapter.

TAke foure or fiue meaſures of the beſt whyte wine, or of ſimple water, or of Maye dewe, or of other lycour pure, accor-

C.ij.   ding

Figure 5. Distillation apparatus, from Conrad Gesner,
The newe iewell of health (1576) [16].

An awareness of the number of editions, and thus the popularity of these books, is crucial to their value as indicators of literacy. Repeated editions indicate that, despite the lack of useful medical information, readers wanted these books and believed them helpful. As Slack noted, the average number of medical works grew from one or two editions (1520-1525) to four or five per year by 1600 [17, p. 240]. Furthermore, these collections of remedies and regimens accounted for one-half of all medical books published during the sixteenth century [17, p. 246]. Designed for ready reference, as was discussed in Chapter 3, these cheap, pocket-sized books accounted for one-eighth of all titles, but one-quarter of all editions published [17, p. 247].

The popularity of these works suggests that they appealed to their target audiences. According to Slack, Elyot's *The Castle of Helth* [18], with its sixteen editions, became the chief representative of the self-help medical "book of secrets" [17, p. 250]. It also became one of three major sources for instruction in nursing [27, p. 49]. In the proheme (proem) of *The Castle*, Elyot stated that he was exposed to the works of Galen, Hippocrates, and other renowned Greek physicians. Although, as he stated, he had not studied at the great medical centers, he believed that his views on healthful living would be of value to all readers. He argued that familiarizing readers with signs of illness enabled the sick to discuss their ailments with their physicians who could then more readily prepare the appropriate medicine.

While Elyot's style in the proheme of *The Castle* echoes the style of *The Boke Named the Gouernour* [28], his renowned book on the education of the ruling classes, the four books of *The Castle* [18] illustrate a prose that is communicative and utilitarian rather than philosophical and ceremonial. As was typical of other technical books, Elyot's style in *The Castle* reflects qualities advocated in modern instrumental discourse—active voice clauses, concrete words, simple or compound clausal units, and concrete description. His segment on the behavior of cucumbers in the stomach illustrates his ability to write concretely and vividly:

Cucumbers.

Cucumbers doe not exceeded so much in moysture as Melons, and therefore they bee not so soone corrupted in the stomache, but in some stomackes, being moderately used, they doe digest well: but if they bee abundantly eaten, or much used, they ingender cold and thick humours in the veines, which neuer or seldome is turned into good bloud, and sometime bringeth in feuers. Also they abate carnalllust. The seeds as well thereof, as of Melons and Gourds, being dried and made cleane from the Huskes are very Medicinable against sicknesses proceeding of heate, also the difficulty or let in pissing: they be cold and moist in the second degree [18, p. 28].

The Castle, as noted above, is subdivided into four books and begins with a table of contents. The first book is composed largely of tables that serve as quick

guides to illnesses and healthful living (see Chapter 2, Figure 1, p. 16; Chapter 3, Figure 13, p. 57). The second book emphasizes descriptions of foods and their positive and negative effects on the body, as illustrated in the preceding excerpt on cucumbers. In the third book Elyot describes common illnesses in some detail. The final book combines descriptions of remedies and conditions, presented in table and sentence form. The general design of the book suggests that Elyot realized that his audience would read selectively (looking for information about a specific condition) rather than analytically.

Elyot's belief that this method of presentation suited his audience and material becomes evident when we examine his most famous work, *The Boke Named the Gouernour* (8 editions, 1531-1580) [28]. Comparing *The Gouernour* and *The Castle* [18], we can see how Elyot's concept of audience and purpose shaped his presentation in each book. In writing *The Castle*, Elyot understood that his readers would come from a variety of educational backgrounds and that they needed simple prescriptions for healthy living. In contrast, Elyot wrote *The Gouernour* for princes and members of the ruling class to suggest the kind of education their children needed to prepare them for a life of public leadership. Anticipating a committed readership, Elyot presented the text in standard undifferentiated prose format. He used a style significantly more verbose and complex than in *The Castle*. Clause units are longer and vary in length from twenty to sixty-five words, although forty words is the most common length. In choosing a style for *The Gouernour*, Elyot knew that 1) his audience was narrower; 2) readers would be culturally and educationally more sophisticated than readers of *The Castle*; 3) this more educated audience would likely read the book carefully and evaluatively in its entirety, rather than as a reference work to help define a malady; and 4) readers would have the reading skill and the interest level to comprehend the complex sentence structures and concepts he used to argue for the appropriate education of England's future leaders.

As we have seen in Chapter 3 (Figures 13, p. 57, and 14, p. 59), content and audience affected typographical presentation. But from a closer perspective, we can see that Elyot made other distinctions based on audience. Elyot opened *The Boke Named the Gouernour* in a heraldic vein that exemplifies the early humanist assimilation of classical texts. He used classical terms for government (*res publica*) and classical examples. Its high ceremonial style, evident in the following excerpt, contrasts sharply with the utilitarian style of *The Castle of Helth* (as exemplified in the previous description of cucumbers):

> A public weale is a body lyuyng copacte or made of sondry astates and degrees of men/whiche is disposed by the ordre of equite and gouerned by the rule and moderation of reason. In the latin tongue hitis called *Respublica* of the which the word *Res* /hath diuers significations/and doth nat only beoken that/that is called a thynge/whiche is distincte from a persone/but also signifieth astate/condition/substance/and profite. In our olde vulgare/psite is called a weale: And it is called a wealthy contraye wherein is all thung that is

profitable: And he is a wealthy man/that is riche in money and substance. Publicke (as Varro saith) is duriuied of people: which in Latin is called *Pupusus*. wherefore hit semeth that men haue ben long abused in calling *Remublica* a comune weale. And they which do suppose it so to be called for that/that every thinge shule be to all men comune without discrepance of any astate of condition/be there to moued more by sensualite/than by any good reason or inclination to humanite. And that it shall sone appaere vnto them that wyll be satisfied either with autorite/or with naturall ordre and example [28, Proheme, pp. 1-2].

In many passages in *The Gouernour*, Elyot's descriptions are fervid or fearful. He states that, when reading Homer "reders shall be so all inflamed, that they most fervently shall desire and coveite, by the imitation of their [the heroes'] vertues, to acquire semblable glorie" [28, Book 1, p. 59]. He notes that grammar, if it is introduced too early in a child's education, will kill "the sparkes of fervent desire of lernynge," which should be the fruit of "the most swete and pleasant redinge of olde autours" [28, Book 1, p. 55]. Even when he wants to reflect on a virtue rather than praise it, Elyot slips from discussion into praise and horror (Placabbilitie [28, Book 2, p. 55]). A virtue is a paradisiacal refuge from monsters [28, Book 2, p. 92]. The style of *The Gouernour* embodies a series of medieval "dicts" amplified by exclamations of wonder, admiration, and despair from a good servant who wishes to inspire devotion and obedience. The contrast between the style of *The Gouernour* and the style of *The Castle of Helth* is striking.

Elyot is one of several writers of both technical and philosophical books who used Ciceronian ceremonial discourse in his philosophical works and a classical plain style in his technical works. The interesting point, however, lies in the differences Elyot himself employed in writing *The Boke Named the Gouernour* and *The Castle of Helth*—clear indication that he saw the need for a utilitarian structure and style for practical prose as opposed to an ornamental prose with minimal partition for his philosophical treatise.

Technical books consistently fell into the category of works using the moderate, low, or utilitarian style, the evolution of which will be the topic of Chapter 5. The interesting point is that technical writers did not see the need for grandiloquence. What we find are sentences with tight clausal structures and common terminology to ensure the clarity of the idea for readers.

General Instruction Books for Daily Tasks

The rude technology available to facilitate common tasks can also be seen in how-to books. These are best described as a conglomeration of various processes that lack a classification system. Examples of this type of plain English instruction for processes include a book of procedures on making gold ink and gold inlay to decorate vellum, parchment, and coats of arms [29] and Leonard Mascall's

A profitable boke declaring dyuers approoued remedies to take out spots and staines [30]. There were books for engravers explaining how to make inks and colors, such *A booke of secrets* [31]. Sir Hugh Platt, an inventive Elizabethan technologist, wrote three books providing instructions for numerous tasks, such as *The iewell house of art and nature*, which contained instructions for distilling plants, preserving fruits and flowers, and methods of molding and casting [32]. The popularity of these books suggests that readers were interested in instructional books on general subjects. The plain style and common diction, reminiscent of spoken English, would appeal to a general audience. For example:

> To temper golde or siluer wherewith you
> may write with a pen, or painte with
> a pencell.

Take fiue or sixe leaues of beaten golde or siluer, and grinde them well and finely on a painters stone with a litle honnye, then put it into a glasse with a quantitie of faire water, and let it stande one night, then drame the water and the honny afterwarde from the golde, and put to the golde gume water, and then write with it, and when it is drye burnith it with an ore tooth. Also if you grinde your leaues of golde with glaire onely without honney putting to it, you may wel write therewith in adding to it a little gumme water, and with your gold tempered in maner abouesaide you may diaper with a small pen or pencell upon colours [31, n.p.].

Cookbooks

Cooking was one technology that developed significantly during the Renaissance, and by the 1580s cookbooks reflected classification and partition as well as expanded recipes for foods. While early sixteenth-century books were disorganized collections of basic recipes, late sixteenth- and early seventeenth-century collections contained tables of contents and divisions of recipes into categories such as "marmalades, conserues, tartuffes, gellies, breads, rock candies, sucket candies, cordiall waters, and conceits in sugar-works" [33].

The extensive classification system used in these small books indicates the state of technological development of cooking as well as the needs of readers and their concomitant knowledge and reading levels. For example, one of the first extant published cookbooks provided only basic recipes for less than two dozen meal items. In contrast, early seventeenth-century recipe books were divided into sections of recipes for breads, jellies, meats, candy, cosmetics, and medicinals. The style, as suggested by the excerpts below, evolved from a telegraphic linear listing of ingredients to a more involved presentation of instructions using complex sentences and more specific quantities. In Pynson's 1508 cookbook, the following meat dish is typical of the recipes included:

for to make chckyns in Musy

To make chckyns in musy / take smale chckyns chopped and boyle theym
in swete broth and wyne and putte therto percely and sage and powder of
peper or graynes and colour it with saffron / then take whyte of egges and ale
drawen through a cloth and put therto and styre it well togeder and put thereto
an unce of gynger and whan it begynneth to boyle set it from the fire and serue
it [34, n.p.].

In contrast, Murrel's *A Daily Exercise for Ladies and Gentlewomen* (5 editions,
1614-1638) provided precise instructions [35]. While the additive style was dominant, the syntax was more sophisticated than that used in Pynson's book of
cookery and exemplified some analysis rather than a speech-based aggregation of
directions. Diction can be characterized as words that would have been used in the
common speaking vocabulary. The detail of these instructions suggests that they
were intended to be read rather than to serve as memory prompts:

Strawberry cakes

Take a quart of very fine flower, eight ounces of fine sugar beaten and cerfed,
twelve ounces of sweet butter, a Nutmegge grated, two or three spoonefuls of
damaske rosewater, worke all these together with your hands as hard as you
can for the space of halfe an houre, then roule it in little round Cakes, about
the thicknesse of three shillings one vpon another, then take a siluer Cup or a
glasse some foure or three inches ouer, and cut the cakes in them, then strow
some flower vpon white papers & lay them vpon them, and bake them in an
Ouen as hotte as for Manchet, set vp your lid till you may tell a hundredth,
then you shall see the white, if any of them rise vp clap them downe with
some cleane thing, and if your Ouen be not too hot set vp your lid again, and
in a quarter of an houre they wil be baked enough, but in any case take heede
your Ouen be not too hot, for they must not looke browne but white, and so
draw them foorth & lay them one vpon another till they bee could, and you
may keep them halfe a year the new baked are best [35, p. 4].

One of the most popular books was *Delightes for Ladies, to adorne their
Persons, Tables, closets and distillatories* (13 editions by 1654) [33]. Other
popular books included John Partridge's *The widdowes Treasure* (9 editions,
1585-1639) [36] and *The treasurie of commodious conceites, and hidden secrets*
(1573-1637, 13 editions) [37], and Thomas Dawson's *The good huswifes Iewell*
(at least 4 editions, 1580-1610) [15]. These books, written in a direct style
suggesting spoken English, provided a rich source of instructions for tasks important to the daily lives of English people. Other books on the preparation of oils
and minerals, many of them translations, provided an avenue other than herbal
preparations for treating diseases. The popularity of these books as well as their
increasingly sophisticated content suggests that women readers were advancing in
their ability to read throughout the Renaissance. Perhaps cookbooks, because the

contents would have been familiar to women, also served as self-help texts by which women could improve their reading skills on their own.

Works on Silkworm Production

Books on silkworm production indicate similar shifts in the knowledge and reading capabilities of intended readers. In the early sixteenth century, propagation of silkworms was mainly a woman's profession. Even though the silkworm industry was not recognized as a guild, because it was female dominated, it was pursued on the lines of the craft guilds of male workers. The silkwomen accepted apprentices, employed workers, conducted their own business transactions, and were sufficiently organized to present petitions to the government [38, pp. 324-335]. These books, when examined chronologically, show the increasing knowledge in the technology of silkworm production as well as increasing reader literacy level.

The earliest English printed works on silkworm production were written in stanza form. T.M.'s *The silkewormes, and their flies* (1 edition, 1591) covers thirty separate topics written entirely in verse [39]. Yet, through verse, the author presented information about propagation, dyeing, the value of the eggs, temperatures for keeping the eggs, time for hatching, collection methods, sickness, metamorphoses from worms into flies, preservation of eggs, and the uses of silken threads. Its ballad stanza suggests that it was written to be read aloud and then memorized by women who may not have been able to read. Because stanzaic presentation had to ensure rhyme, the precision and detail of procedures diminished:

> The first three weekes the tend'rest leaues are best,
> The next, they craue them of a greater size,
> The last, the hardest ones they can digest,
> As strength with age increasing doth arise:
> After which time all meate they do detest,
> Listing vp heads, and feete, and breast to skies,
>   Begging as t'were of God and man some shrowde,
>   Wherein to worke and hang their golden clowde.
>
> But whilst they feede, letal their foode be drie
> And pull'd when *Phoebus* fact doth brightly shine,
> For raine, mist, dewer, and spittings of the skie,
> Haue beene ful of the baine of cattle mine:
> Stay, therfore, stay, til dayes-vpholder flie,
> Fiue stages ful from Easterne Thetit line:
>   Then leaues are free from any poysned feede,
>   Which may infect this white and tender breede [39, p. 55].

A second work, Stallenge's *Instrvctions for the increasing of Mulberie trees, and the breeding of Silke-wormes, for the making of Silke in this Kingdome*

(2 editions, 1609) [40], was published two decades after T.M.'s work. This little book, equally brief (13 pages), exemplifies the shift to prose to convey instructions. Comparing the following excerpt with that from T.M.'s work reveals how prose was emerging as the preferred vehicle for explaining the silkworm processes. The increased complexity of the style also illustrates that a more complex knowledge of silkworm production had emerged. The difference in styles further suggests that the literacy of readers interested in silkworm production, likely the silkwomen, had increased.

What ground is fit for the Mulberrie seedes, how the same
is to be ordered, & in what sort the seedes
are to be sewed therein.

The ground which ought to be appointed for this purpose, besides the naturall goodnesse of it, must be reasonably well dunged, and withall so scituated, as that the heate of the sunny may cheris it, and the nipping blastes of either the North winde or the East, may not annoy it: the choice thereof thus made, that the seeds may the better prosper, & comb vp after they bere sowne, you shal dig it two foote deepe, breaking the clods as small as may be, and afteward you shall deuide the same into seuerall Beds of not aboue fiue foote in breadth, so that you shal not neede to indanger the plants by treading vpon them, when either you water or weeded them [40, p. B2].

A third book, DeSerres's *The perfect vse of silk-wormes and their benefit* (2 editions, 1607-1609) [41], was one of the silkworm books read by the housewife [42, p. 44], but it also approached silkworm production as a major industry. Approximately 100 pages long, this work provided much more extensive as well as detailed explanation and instructions for growing and harvesting silkworms. It included some of the same visuals used in the earlier work on silkworm production, but the extent of detail in describing the process indicates that 1) knowledge of silkworm production had indeed taken a gigantic leap and 2) the literacy level of readers had also increased:

To gather the leaues for to be giuen to the Wormes.

For the order which one is to hold in gathering the mulberry leaues, for the victuales of these creatures, consisteth the second article of this work, for to make the trees of a perpetuall seruice. It is to be noted, that to plucke off the leaues bringes great damage to al trees, oftentimes euen causing them to damage to al trees, oftentimes euen causinng them to dye: but seeing that the Mulberry is detined to that, naturally supporteth such tempest better than any other plant: yet neuerthelesse you must goe to it very retentiuely, for to disleaue the Mulberrie inconsideratlie is the way to scorch them for euer, to cause them miserably to die in languishment. Euery one confesseth that to gather the leaues with both hands, leafe after leafe, without touching the shoote, is the most assured way for conseruation of the trees [41, p. 25].

The popularity of these books tells us a great deal about not only the reading comprehension and hence the literacy of English readers, but also the interests of ordinary English people, the work they had to perform as part of their daily lives, and the increases in knowledge in medicine, in cooking, and in fields such as silkworm production. The consistency of presentation in books written for men, for women, and for both men and women suggests that women were becoming as literate as men.

TECHNICAL BOOKS CONTAINING SPECIFIC SECTIONS FOR DIFFERENT AUDIENCES

Sensitivity to an audience's reading needs and interests can also be found in books that contained sections written for men and sections written for women. Most of the books in this category dealt with farming, gardening, and estate management, topics needed by both male and female readers of the merchant classes. As Wright noted, "the writings of Fitzherbert, Tusser, and Markham long remained popular because London merchants, having made sufficient money to buy a country estate, bought these books to help them manage their new purchase and lead a successful rural life" [2, p. 565].

Books on Estate Management

The oldest and most popular book on husbandry and household management of the English Renaissance was Thomas Tusser's *Fiue hundreth points of good husbandry* [43]. This work, written completely in verse, enjoyed twenty-one editions from 1557 to 1607 and included sections on farming management and "The Points of Huswiferie." Agricultural historians assume that this earliest printed farming manual was popular because it could be easily read by those with minimal reading skills and read aloud to and memorized by people having no reading skills. Tusser's book uses mostly folklore-based instructions which remained unchanged through its twenty-one editions.

Books on farming and estate management by Gervase Markham, written six decades later, also contained sections for men and women [44, 45]. Markham's books, characterized by their detailed process descriptions and instructions, exemplify modern technical style. The large number of topics covered and the increased information about each topic, compared with Tusser's book, shows that technology in farming had advanced rapidly between 1557 and 1614. While Tusser's work remained popular, the popularity of Markham's books suggests that readers had emerged who now wanted a more sophisticated approach to agriculture. Yet, the concomitant popularity of both Markham's works and Tusser's book indicates that literacy levels among readers interested in farming were still uneven.

Markham's *Countrey Contentments* (5 editions, 1614-1633) contained two books, one for men and one for women [44]. Both had detailed tables of contents. Book I, "The Husbandmans Recreations," dealt with methods of caring for hounds; instructions for hunting hares and deer; instructions for breeding, training, and riding horses; instructions for raising and caring for hawks and greyhounds; instructions for shooting the longbow and crossbow; and instructions for bowling (croquet) and for tennis. Book II, "The English Hus-wife," included remedies for various illnesses; instructions for kitchen gardens; recipes for salads, breads, meats, herb dishes, sauces, desserts, pastries, and jellies; instructions for dyeing wool and spinning; instructions for growing flax and preparing it for spinning; instructions for dairying; and information on brewing beer and ale.

"The English Hus-wife" differed from other books on similar topics, such as Tusser's, in that it provided more informative explanations. Markham apparently assumed that female readers who had recently moved to the country needed substantial process descriptions and instructions on some topics. Thus the chapters on wool, hemp, flax, clothmaking, dairying, and brewing provide more explanation than the chapters on cookery and "physick," topics that city wives would have already known about. A comparison of the content development of each section suggests that Markham was sensitive to the level of detail that women readers needed.

The style Markham used throughout "The English Hus-wife" does not suggest that he believed that female readers were any less literate than male readers. Excerpts from Book I and Book II of *Countrey Contentments* illustrate the similar styles:

> Touching the ordering of Hawkes, the first thing the Faulconer must doe to his Hawke after shee is taken from the Caidge, is to bathe her in warme water and pepper, beeing no more but luke-warme, thereby to cleanse her from nits, lice, and such like vermine; then he must for her generall feeding rather keepe her hie in flesh then poore, because thereby he auideth disease, then after euery flight whether it be at pray, at traine, or at the lure, he shall giue his Hawke casting, if she bee a long-winged Hawke hee shall giue her flannel [44, Book I, Chapter VIII, p. 88].

> To make dry vinegar.

> To make drie Vinegar which you may carrie in your pocket, you shall take the blades of greene corne either Wheat or Rie, and beate it in a morter with the strongest Vinegar you can get till it come to a past; then role it into little balls and drie it in the sunne till it bee very hard, then when you haue any occasion to vse it cut a litle peece therof and dissolue it in wine and it will make a strong Vinegar [44, Book II, Chapter II, p. 82].

This similarity is significant for two reasons: 1) "The English Hus-wife" also appeared in at least three separate editions, which suggests the continuing popularity of the work among women readers, and 2) Markham, one of the most

prolific writers of the Renaissance, showed a remarkable ability to write romances and histories as well as how-to books. Markham's book on husbandry, written for serious farm managers, displays a highly succinct style and a content formation that focuses on animal propagation rather than on the care of sporting animals, which would have appealed to gentlemen's interests [45]. Thus, each kind of work, prepared for readers with different interests, exhibits a different style, which suggests that Markham realized that different readers and different prose genres *required* different styles.

In contrast, when Markham tried his hand at prose dream romance, as in *The English Arcadia* (3 editions) [46], he chose a cumulative sentence structure that emphasizes languid, rather than concise clauses; characters from Greek myth; and an almost lyrical connective phrasing not suitable for referential reading but for slow, evaluative reading requiring readers to hold in memory each part of the cumulating sentence. The style itself anticipates a stream-of-consciousness form that captures the purpose of the work—conveying the dream vision of the narrator in a language that allows the reader to share the experience:

> The infinite varieties wherewith the hand of Fortune feedeth the hungry eares of change-desiring-man; are so full of hony-poysons, that with our uncloid appetites wee seeke to swallow that with delight, which with greatest earnest-nesse wee haue fled from, and eschew'd as our worst torment; al-bew the face crry euer one character, onely the shape, disguid'd in a roabe of new fashion, as (most excellent sonne) thou shalt perceiue by this continuance of my most true History. For no sooner had the Maiden, wonder of all beauties, (I meane the incomparable Princesse *Mellidora*) and the sharpe witted *Ethera*, con-ueyed my drowsie-dead-seeming body into a priuate Arbor, adioyning vento the Chappell of the great God *Pan*, but there, they two, with there dainty hands, disrob'sw me of my gowne, of my hood, of my booke, of my beads, of my glasse, and of my staffe [46, n.p.].

In contrast, in his estate management books, such as *Countrey Contentments* [44] and *Cheape and Good Hvsbandry* [45], Markham uses an efficient subject-verb-object arrangement within active-voice clauses (see Chapter 3, Figure 27, p. 81, for a sample page from the latter book). In his husbandry books, a sharp, denotative style replaces the liquid, allusive style of the dream vision.

Another estate management work, a clear precursor of Markham's, was Fitzherbert's *Booke of Husbandrie* (11 editions, 1526-1598) [47]. This work grew in number of topics and extent of coverage on each topic, but the basic style was unchanged throughout its eleven editions. The difference in content of the sections for men and the sections for women suggests the differences between men's and women's roles, but it also suggests that Fitzherbert saw that his male and female readers were equally literate and that the information they needed for performing their expected tasks was the same. Compare the following two excerpts, the first addressed to women and the second to men:

Chapter 6.
Howe to choose the best Hennes for broode.

If you desire to make choyse of the best hennes for broode, you must in all poynts haue them of the same colours which I haue already shewed in the choyse of your Cocks, though they neede not bee eyther so hie, or big of body. They must be large brested and bigge headed, hauling a straight redde double comme, great white eares, and her tallons suen. The best kinde are such as haue fiue clawes, so that they be without spurres: for such as haue spurres will yield ye small profit, by reason that with theyr spurres when they sit they breake theyr egges [47, Book IV, p. 149].

Chapter 20.
Of the diuers kinds of Manure, and which is the best.

There be diuers sorts of Manures, and first of those that bee worst, as Swines dunge, which Manure breedeth and bringeth vp thistles, the scourings of Hay bsarnes, or Corne barnes, which bringeth vp sundry weedes and quirks, and rotten Chasse, which diuers vse, but brings little good. The shoueling of Highwayes and streets is very good, chiefly for Barley. Horse dunge is reasonable. The dunge of all maner of Cattel that chew the cudde is most excellent. Doues dunge for colde ground is best of all, but it must bne spread very thinne [47, Book I, p. 28].

Books on Gardening

Technical books are important sources for showing us the major vocational interests of Renaissance English people, particularly the daily work that had to be done. Renaissance England's dependence on agriculture, which required that both men and women assume active roles to generate sufficient food, explains why books in the general agriculture category contained specific sections for men and for women. These works partition tasks according to gender. Books on gardening, like agriculture books, also included sections for men and for women. While woodcut drawings in these books suggest that men and women often worked side by side, tasks usually done only by men required more physical strength, while women were usually assigned the propagation and harvesting of the kitchen garden. William Lawson was one of the main writers of gardening books. Lawson's works clearly show that he was aware of the content needs of readers, and his works were directed toward readers with different gardening tasks.

Lawson's *A New Orchard and Garden* (5 editions, 1617-1637) contained two parts, one for men and one for women [48]. Its content emphasized that gardening tasks were assigned according to gender, as the work's subtitle suggests: "The best way for planting, grafting, and to make any groud good, for a rich Orchard. . . . With the Country Housewifes Garden for hearbes of common vse, their vertues seasons, profits, armaments, variety of knots, mydals for trees and plots for the best ordering of Groundes and Walks." Lawson's sections directed to men show

his awareness that men likely had received Latin instruction in grammar schools, as we find an occasional Latin phrase. His sections for men are longer and more detailed than parallel sections for women. However, a comparison of the content of the sections for men with those for women shows that Lawson expected female readers to be as literate in vernacular English as male readers. The lack of detail and lesser content in the sections for women suggest that women's work was less arduous than men's because of physical limitations rather than intellectual ones. Women had many of the same chores as men, but the size of the garden limited their work and hence the knowledge of common tasks required of women gardeners. For example, Lawson devoted over two pages to Chapter 5, describing the form of the garden. The following excerpt echoes the style used throughout the opening part written for men:

> The goodnesse of the Soyle, and Site, are necessarie to the well being of an Orchard simplie, but the forme is so far necessarie, as the owner shall think meete for that kinde of forme wherewith euery particular man is delighted, we leaue it to himselfe, suumcuig, pulchrum. The forme that men like in gnerall is a square, for though roundnesse, be forme perfectissims, yet the principle is good where necessitie by art doth not force forme other forme. Now for as much as one principiyall end of Orchards is recreation by walks, and universallie walks are streight, it followes that the best forme must be square, as best agreeing with stright walks: yet if any man be rather delighted with some other forme, or if he ground will not beare a square, I diseommend not any forme, so it be formall [48, p. 12].

In the section on form, Lawson included a labeled drawing (see Figure 6), a Renaissance technical description of the form of an orchard and garden with mazes. "The Couentrey Housewifes Garden," included as Part II of *A New Orchard and Garden*, was much shorter than Part I (23 pages compared with 57 pages). Each section was limited to a page or perhaps only a short paragraph. Some topics were discussed in both Part I and Part II, such as "The Soyle," "The Site," "The Forme." The sections in the housewife's part were less detailed than those in Part I. For example, "Chapt. III. Of the Forme" was given only one paragraph:

> Let that which is said in the Orchards from suffice for a Garden in generall: but for speciall formes in squares, they are many, as there are devices in Gardiners braines. Neither is the wit and art of a skilfull Gardiner in this skillful point not to be commended, that can worke more variety for breeding of more delightsome choise, and of all those things, where the owner is able and bestrons to be satisfied. The number of formes, Mazes and knots is so great, and men are so diuersly delighted, that I leaue euery housewife to her selfe, especially seeing to set downe many, had been but ti fil much paper; yet left depriue her of all delight and direction, let her view these few, choise, new formes, and note this generally, that all plots are square, and al are bordred

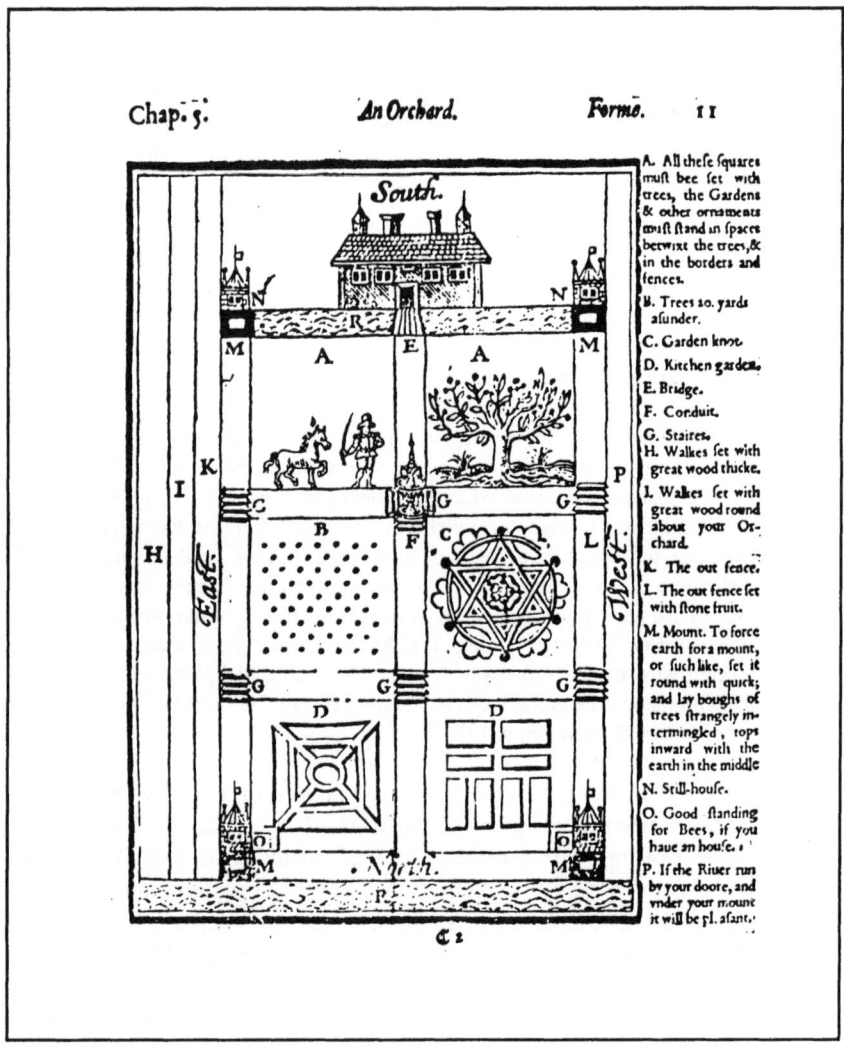

Figure 6. Maze configuration in section written for men (Part I), from William Lawson, *A New Orchard and Garden* (1623) [48].

about with Pruit, Raisns, seaberries, Roses, Thorne, Rosemarie, Bee-flowers, Fsop, Sage, or such like [48, p. 2].

Figure 7 gives configurations for mazes from Part II. Comparing Figures 6 and 7 suggests that women were in charge of designing and planting the mazes within the larger garden, which was the male gardener's province.

Figure 7. Maze configuration in section written for women (Part II), from William Lawson, *A New Orchard and Garden* (1623) [48].

Lawson's sensitivity to differences among readers can be seen further when we compare Thomas Hill's *The Gardeners Labyrinth* [49] with *A New Orchard and Garden* [48]. *The Gardeners Labyrinth*, with its extensive treatment of gardening and highly detailed discussion of gardening methods using horticultural nomenclature, suggests that gardening was a serious vocation. Professional gardeners demanded a more rigorous treatment of gardening instructions and process descriptions than did gardeners targeted by *A New Orchard and Garden*, which limited its coverage to discussions of broad topics such as the site, the soil, and the form of the garden. A glance at a segment of the Table of Contents of *The Gardeners Labyrinth* (see Figure 8) suggests the extensiveness of the treatment of gardening in this work compared with *A New Orchard and Garden* (see Figure 9).

TECHNICAL BOOKS FOR EXPERT READERS

The fourth category of technical books is those written for expert readers. Not as large a market existed for these books as for books that provided more general instructions for less skilled readers. Popular books for expert readers were unlikely to have appeared in more than one or two editions. Comparing the level of expertise offered by a wide range of books helps us understand the state of knowledge in many technical fields and the range of literacy among readers interested in these fields. As we might expect, the best examples of technical books for expert readers can be seen in the medical and surgical books written for practicing physicians and surgeons, navigation books written for sailors, military science books for officers, and a few books for expert gardeners.

Books on Surgery and Medicine

As we examine books for physicians, we should recall the content, style, and presentation of many of the self-help medical books, which used layman's terminology to describe physical maladies. We find contrasts among the uroscopy books. For example, *Here Beginneth the Seynge of Urynes* [26] uses common language to describe urine color and density, while *The Differences Cavses and Ivdgements of Vrine . . .* [50], a more sophisticated uroscopy for physicians, uses medical phrases, such as "Obstructions and stoppings of the emulgent veines," "Cruditie and flowe concoction," "causes which doe ottenuate and make thinner any matters in the bodie." The initial section on deformed contents found in the urine appears as follows:

> Of deformed contents, altogether unnaturall, and alwayes ill; which appeare either in the sediment or lowest region, or in the swimme and middle region, or lastly in the cloud or upper region. And first of unnaturall and deformed contents in the sediment, which are commonly reckoned fifteene:

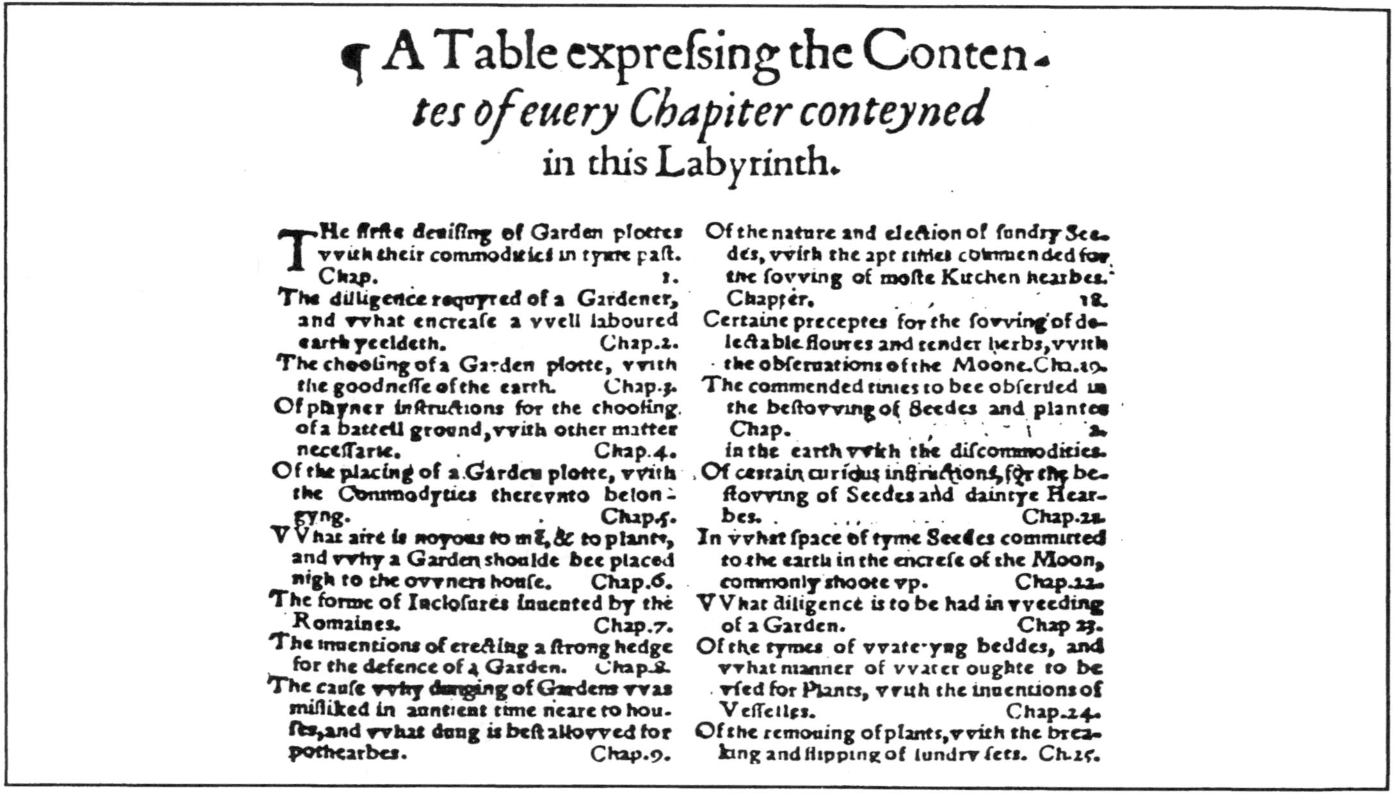

¶ A Table expresſing the Contentes of euery Chapiter conteyned in this Labyrinth.

Figure 8. Excerpt from the table of contents, from Thomas Hill, *The Gardeners Labyrinth* (1577) [49].

The Contents.

Figure 9. Excerpt from the table of contents, from William Lawson, *A New Orchard and Garden* (1623) [48].

1. Popinjay greene or oylie sediment signifieh.
   Colliquation of the whole body, with paine, and ague, and pissed flow by little and little: or els colloquation of the reins or bladder onely with paine in the reines, no ague, and pissed fast.
2. Spider webe or fatnesse sweiming in the urine, signifieth Consumption of the whole body with an gue.
3. Red vetches or fitches, ervaceum, orcbeum, signifie
4. Great inflammation of the liver.
5. Colliquation of the whole body: or of the reines onely, with conditions as before in popinjay greene.
6. Plates, scales, (folium, lamine, squame) hauing breadth and length onely, like scales of fishes, especially a gogeon [50].

As previously discussed (and shown in Figure 2), lack of English equivalents for many Latin medical terms led to increasing use of Latin terms (as shown above in items 3 and 6), particularly when no equivalent English term was available.

Books on surgery were written with the surgeon as the specific audience. The purpose of these books was to describe surgical instruments and procedures for their use. As I will discuss in Chapter 6, these books contained the best examples of English Renaissance technical description and incorporated terminology appropriate to the discipline. William Clowes, a renowned English military surgeon, described methods of treating gunshot wounds and venereal disease in *A Profitable and Necessarie Booke of Obseruations* . . . [14]. Clowes used a case-study approach, a first-person narration of his experience treating a specific wound, and included pharmaceutical preparations for treating gangrene, infection, bruises, etc. Despite the uniqueness of his treatment for wounds, based on his experience with patients, Clowes's treatment of venereal disease showed his reliance on Galenic medicine and astrology. For example, his instructions for bloodletting begin as follows:

> Concerning letting of blood, which is the second kind of evaucation: I hope it may be lawfull for me in some sort to speak, according to mine owne experience and obseruations herein.
>
> It is commonly called the opening of a veine, wherein are many obseruations to be noted, especially these three: the strength of the body, the constitution of aire, and the motion and place of the planets [14, p. 159].

Basically, books for experts differed from works for lay readers in the extensiveness of the information provided, the sophisticated level of treatment suggested, and even the medical problems discussed. Most treatments prescribed (which required a knowledge of pharmacology) would have been beyond the capability of medically uneducated readers. Self-help medical books generally dealt with simple remedies for common problems, such as nosebleeds, sore throats, and chills. Books for physicians and surgeons provided detailed discussion on using cautery tools for stopping nosebleeds, on initiating bloodletting for curing chills and fevers, and on procedures for excising tumors and setting broken

bones, techniques that non-medically trained readers would have lacked the equipment, the pharmaceutical ingredients, and the skills to employ.

Clowes, for example, provided extensive (though certainly neither scientific nor accurate) descriptions of the causes of venereal disease, and then the cures—medicinal preparations, bloodletting, and diet [14]. These sections precede detailed case studies of patients Clowes had observed who had been treated according to these suggested remedies.

John Banister's *A Needefull, new, and necessarie treatise of Chyrurgerie* . . . focused on excision and topical treatment of ulcers [13]. Latin occurs in the marginal headings and in pharmaceutical specifications. Banister, like many English physicians, borrowed heavily from Latin medical nomenclature when English equivalents were unavailable. The opening lines "Of the virulent corroding and seding Vlcer" show this combination:

> These kind of vlcers which differ only Sedcundum maium & minus, must haue the Methode of their cure consiste in three thinges onely. Firste in a diet colde and drie, secondly by euacation of the antecedent matter, both by Phlebotomie if it seme profitable as also by medicines exiccatiue, as is sayde in Herpes, and Erisepelas. Also particuler reuulsions are comodius, as cupping, frication, and binding the corrarie member, of use of vng. de Bolo, to beate backe the matter flowing. The thirde scope is referred to medicines colde and drie astringent and discutient [13, p. 27].

Another type of book for practitioners was the surgical instruments book, which described surgical tools (usually with drawings) and the use of these tools in surgical treatments for maladies such as harelip, skull fractures, and hernias. A translation of Paré's *An Explanation of the Fashion and Vse of Three and Fifty Instruments of Chirvrgery* (1634) was one such book [51]. This large folio includes technical descriptions of needles, trusses, trepans, drills, and scalpels. Figure 10 shows a page from this work that describes the treatment for harelip. The writer assumed that the reader would have had exposure to anatomy, probably a version of Vesalian anatomy, and would be familiar with terms such as abdomen, peritoneum, cicatrice, and omentum. However, even if vernacular readers were unable to read Latin, they would still be able to understand the main thrust of the treatment described.

Works on Navigation and Geography

Another major topic in Renaissance technical writing concerned navigation and geography, and these books were also written for either specific or general audiences. English Renaissance readers purchased the most precise, descriptive works, as can be seen in the differences in numbers of editions. The popularity of the more detailed books suggests the rapidity with which literacy was increasing. Elizabethan naval conquests—in addition to the growing literacy and an increasing interest in news of foreign lands and cultures, potential investment

breach,and halfe at the other; and in the same manner, another short needle (as the place will beare)must be thrust in the middest, crosse the former, and vnder the flesh as the former is. Then the thred which was in the eye of the first and longest needle, must be strayned crosse about all the foure ends of the two needles, seuen or eight times, or more, to hold the lips of the wound or breach firmely together till the vnion be perfected, and then the needles must be taken away, and the wounds they make healed according to Art. But if it be a Hare-lip that was a fault of Conformation, the skin on both sides of the diuision must first be cut away, otherwise the sides will not vnite. The figure of this kinde of Seame you haue hereunder expressed.

The forme of the Suture for the Clouen or Hare-lip, together with the needle and thred rowled about them.

The fourth kind of Suture is called by a proper name *Gastroraphia* : and is proper to the lower Belly, when the muscles of the *Abdomen* and the *Peritoneum*, are mainely wounded; in so much that some of the entrals sometimes fall out, but the *Omentum* alwayes, at least a part of it. If the wound passe no further then to the *Peritoneum*, it may be cured after the ordinary way of curing simple wounds : but if the *Peritoneum* be wounded, then the worke must be performed after this manner. First with your needle take vp the *Peritoneum* alone, for example. on the right side of the wound, (for you may begin on which you please) and passing by the left side of the *Peritoneu* take vp the flesh of the muscle,& the skin with it on the left side. Then take vp the left side of the *Peritoneum*, and passing by his right side, take vp the right side of the muscle and skinne. Then take vp the right side of the *Peritoneum*, and passing by his left side, take vp the left side of the flesh and the skinne, and so few your Seame crosse in and out, till the whole wound be seamed. The reason of this manner of seame is, because the *Peritoneum*, being a membranous part, and without blood, cannot re-vnite, vnlesse it haue a bloody part ioyned to it, which in this manner of suture it hath quite through the length of the wound. You must also obserue, that this seame must be sewed as close as well may be throughout the length, till you come to the decliue or lower part of the wound, there you must leaue a stitch at least, both in the membrane and in the muscle, at which vent the *sanies* and matter may issue, till there be none left, and then it must be healed vp with a *Cicatrice*. Obserue also, that if any of the *omentum* issue out of the wound, it must not be reposed, but cut off; for that fat is foeculent, and if it once take the aire, or rather the aire take it, it will putrifie. Also if any of the entrals issue at the wound, the Patient must be laid in such a position as they may best be reposed, and that must be done with great tendernesse and easinesse of hand, that the ayre come as little to them as may be, and that they be not violated or shuffled in the reposition.

The fift kinde is called the dry Seame, which is vsed onely in wounds of the face, to auoyd scarres which will make the face deformed. For that is the market place, especially in women, to please whom Chirurgians haue deuised this kinde of Suture.

According to the bignesse of the wound, you must cut out a paire of Indentures, of strong Cloath, in the forme hereunder described. These you shall spread with the following Medicine, or steepe them in it. *Recipe pulerum Mastich ; sanguinis draconis, thuris, farina volatilis, Tragacantha contusa, gypsi, picis, sarcocolla, ana dra. 1, picis nigra scruples 2, grain 5. albumina ouorum qua sufficiant, fiat medicamentum.* These cloathes you shall lay vpon the face, at the opposite lips of the wound, so that they may be distant the breadth of a finger, and so suffer them to lie till they cleaue strongly to the skinne of the face. Then with a needle and thred, straine the indented opposite corners together so hard, that the flesh (to which they will firmely cleaue) may be drawne together with them, and so be revoited by such meanes as you shall finde to be fit. But be carefull with fit bands and tyes about the head to establish your worke, that it stirre not till you haue brought it to perfection. The fashion of the Indentures and their application is hereunder expressed.

h *The*

Figure 10. Surgical procedures for harelip, from Ambroise Paré, *An Explanation of the Fashion and Vse of Three and Fifty Instruments of Chirvrgery* (1634) [51].

opportunities in foreign trade, and exploration and sea travel—stimulated demands from seamen, merchants, and the general reading public for books on geography and navigation. George Abbott's *A Brief Description of the Whole Worlde* (11 editions, 1599-1636) was one of the most popular geography books to appear before 1640 and was apparently popular with seamen preparing to sail outside English waters [52]. The book provided geographical descriptions of thirty-four countries and areas—such as Spain, France, Russia, China, India, Chaldea, Asia Minor, Egypt, and Africa—in approximately seventy pages, and stands as one of the first geography books. Abbott used spatial arrangement to present his description of each country so as to allow the mind's eye to visualize the location of each country before presenting other details about it:

De Italia

On the South-side of the Alpes, and Germanie: lyeth Italie, stretching it selfe out in length toward the South, and East. It hath on the South-side, the Iland Sicilia: on the East, that part of the Med, which is called Mare Adriaticum, or Mare superum; which seuereth Italy from Graecia on the West-side that parte of the Med, which is called Mare Tyrrhenum, or Mare inferum: and by some, Marre Ligusticum.

This countrie for the figure thereof, is by some likened vnto a long leafe of a tree: it hath in the middle of it which goeth all in length, a mightie mountaine, named Mont Apennius, which is likened vnto the Spina, or Ridgebone of the backe: Out of this hill springeth diuers riuers which runne on both side of it, into the Adriatike, and Tyrrhene, or Tuscane seas.

The North parte of this Italy, is that which in antient time was called Gallia Comata, or Gallia Cisalpina, inhabited then by the French-men. It is now called Longobardia, or Lombardie: wherein stand many ritch gouernemets: as the Dukedome of Millaine, or Mantua, of Florence, and others. It is for the pleasantnes thereof, in respect of the soyle, ayre, waters, and great verietie of wines, and fruites: likened now by some, to Paradise, or the garden of God [52].

That an increasing number of English readers were rejecting the medieval mind-set can be seen in the decreasing popularity of geography books that were little more than folklore. For example, *Mappa Mundi* (1535?) [53] and Ierome Turler's *The Traveiler* (1575) [54] provided mythological descriptions of locales, such as the Mediterranean. The writer of *Mappa Mundi*, for example, stated that heaven and hell were located on parts of the earth:

And Paradyse terrester is in the fyrst regylyon of *Allve* logrande toward Orient, and there is no earthly man that may come there but god on his angles lede him and gyde him thyther. for this paradyse for them and Adama was cast and driven out ther by an Angell, anone forthureth it was environed al about with great fyre, that dureth from the erth vnto the skye aboue, that no man may entre nore come therto, And wyldenesse where there is also great multitude of many dyuers & peryllous beestes [53].

Navigation books were generally more popular than geography books, even though they were written for a more knowledgeable audience, the working seaman. The technical development of navigation books during the last quarter of the Renaissance attests to the interest of Englishmen in commerce and exploration and the effects of instrumentation on navigation. Books for seamen ranged from those describing instruments—such as the dial, the pantometer, and the sphere—to books on how to sail to destinations such as Normandy, France, and Scotland.

Richard Eden's *The Arte of Nauigation* included extensive drawings, astronomical diagrams, and tables giving declinations of the sun [55]. Yet, the work's dependence on Greek thought was evident in its definition of creatures, the world, the motions of the heavens, and the elements, and the operation of the zodiac circle. The book also included instructions for making a mariner's compass and how to use the astrolabe. (Figure 11 shows Eden's description of a solar eclipse.)

Similar works included William Bourne's *A regiment for the sea* (12 editions, 1574-1638) [56]; *The Rutter of the Sea* (7 editions, 1528-1573) [57]; *The Mariners mirrour* (1 edition, 1588) [58]; and *The Safegard of Sailers* (9 editions, 1581-1634) [59]. Earlier works, such as *The Rutter of the Sea* [57], used no visuals, no introduction, while launching immediately into instructions for sailing into various harbors. The author assumed that his seamen readers would need no orientation to the descriptions and instructions he was providing. For example, the opening chapter begins as follows:

> Of the tydes, that is to Nort
> the fluds and Ebs, fro the race
> of Sayne in Flaunders.

At the race of Sayne, the mone in the southwest and poynt of south full sea, and in the foutheast and a point of east low water.

At Saint Mathewes the Moone northeast, ful sea, and southweast low water.

At the fourne the moone northeast a point of east ful sea. & in the southwest, a poynt of westlow water.

At Portal and at Bergroath the moone in the east northeast ful sea, Y in the southweast low water.

At Lyganan the moone in the east northeast ful sea, and southeast lowe water.

Wythin the Ile of Basepole the moone in the east northeast full sea, and without the chanel, the moone in the east, and the moone in the south, low water [57].

Later works, such as *A regiment for the sea* [56] and *The Light of Navigation* [60], provided extensive visual aids, more specific instructions on navigation, and detailed maps of shorelines that supported the instructions for approaching each particular shore. Later Renaissance books were better organized than *The Rutter of the Sea* [57] but assumed that readers would be familiar with the terminology

of the Arte of Nauigation.   Fol 35,

although the Eclipse of the Sunne shalbe totall or perticuler, it
can not be vniuersal in the whole earth.

   And note, that for the quantitie of these Eclipses, the Astro-
nomers deuide into twelue equal partes, aswel the Diameter of
the Sunne, as of the Moone, and these partes they call figures,
punctes, or prickes, and accordyng to the punctes of the Diameter of the Moone, whiche is couered by the shadowe of the
earth, or the partes of the Diameter of the Sunne, whiche the
Moone doth couer, so many fingers or punctes shalbe sayd to be
Eclipsed. As yf. 6. the halfe, yf. 3. a quarter, yf. 4. a teree, or
thyrde parte, yf. 6. three quarters, yf.
8. two terees.

*Of the quan-
titie of the
Eclipses*

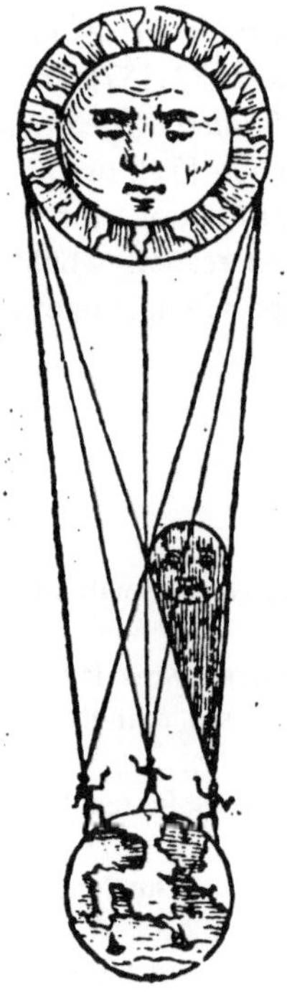

   It is also to be noted, that although
the Sunne be bygger then the Moone,
yet at some tyme the Moone seemeth
greater then the Sunne. And this
shalbe, when the Sunne is in the
Auge of the Eccentrike, & the Moone
in the opposite of the Auge of the
Epicicle.

*Why the
Moone see-
meth somtime
bygger, and
sometime lesse
then the
Sunne.*

   And when it so appeareth, he may
be al Eclipsed. Sometymes also the
Moone seemeth lesse. This is when
the Sunne is in the opposite of the
Auge of the Eccentricke, and the
Moone in the Auge of the Epicicle.
Then although we shoulde see the center of the Moone in the center of the
Sunne, she can not hide hym al whol-
lye, because the Sunne shal appeare
greater.

   Of this that we haue sayd, it foloweth that al the Eclipses of the Sunne,
must of necessitie be in the coniunction, and the Eclipses of the Moone in
the opposition: whereby is inferred,
that the Eclipse of the Sunne in
    C. iii.   the

*The Sunne is
Eclipsed in
coniunction, &
the Moone in
oppolition.*

Figure 11. Description of an eclipse of the sun, from Richard Eden,
The Arte of Nauigation (1589) [55].

and techniques of navigation. Many of these books included instructions for making and using navigational instruments, such as cross-staffs and spheres, and descriptions of heavenly phenomena.

However, even works on navigation could fail to gain popularity if they did not consider the needs of the audience. For example, in 1599 William Cunningham wrote *The Cosmographical Glasse, Conteinyng the Pleasant Principles of Cosmographie, Geographie, Hydrographie, or Navigation* [61]. Too broad in coverage and too general in its treatment of procedures, the book had little appeal to seamen, who responded more favorably to works that provided useful, precise detail, such as *The Light of Navigation* [60] and *The Safegard of Sailers* [59]. Cunningham used a dialogue to discuss cosmography, rather than a well-organized format which featured readily accessible information. However, with its charts and tables, the work was too technical for the non-expert reader interested in descriptions of sea travel.

Agricultural Books

Agricultural works included specialized books, such as Mascall's books on poultry, cattle, and grafting [62-64]. Markham wrote an advanced book on husbandry directed to readers who were more than armchair farmers [45]. The main difference between Markham's *Cheape and Good Hvsbandry* [45] and his *Countrey Contentments* [44], as has already been mentioned, was in quantity and depth of information. *Countrey Contentments* emphasized the care and maintenance of hounds, horses, and falcons—main sources of recreation for country gentlemen; but *Cheape and Good Hvsbandry* prescribes the care and maintenance of animals for more than sporting purposes. The book is divided into a number of sections—horses; bulls, cows, calves, and oxen; sheep; goats; swine; conies; and poultry, which he subdivided into geese, turkeys, waterfowl, and hawks. He concluded with a short section on bees and one on fishponds. Markham directed this book to vocational farmers who wanted state-of-the-art help in raising and caring for a large farming operation (see Chapter 3, Figure 27, p. 81, for a page from this work).

Oliver DeSerres's *The perfect vse of silk-wormes . . .* (1607), already discussed as an indicator of changing literacy and technology during the Renaissance, was the most complete book on raising silkworms published in the English Renaissance [41]. The content of the book suggests that it was written not for the silkwoman who raised silkworms as a small operation, but for major silkworm producers, usually men who transformed silkworm production from a cottage to a capital industry.

For example, DeSerres was very specific about food requirements for silkworms:

> Commonly, a thousand pounds of the leaues of Mulberries being ten
> hundred weight, is sufficient to satisfied and feede an ounce of the seede of

> Silkwormes; and the ounce of graine, makes fiue, or sixe pounds of silke;
> euery pound being worth two or three crownes, and more; wherefore, ten or
> twelue crownes come of ten hundred waite of leaues [41, pp. 11-12].

Labor requirements were also stated:

> And for to partricularise the expenses, I may say, that an hundreth of sixskore
> gathers, wherof three quarters, are women, or boyes, are sufficient to gather
> all the laues necessarie to feed ten ounces of the feed of the the Wormes, and
> to bring them into the place of the cattell, the Mulberries being not farre
> distant from the house as is requisit. To the payment of which worke for the
> qualitie of the persons, ariseth not to much money. For it is in victuals that the
> most is consumed [41, p. 13].

In comparison to the poetic instructions for raising silkworms quoted earlier,
DeSerres's book shows more specific description and instruction. Lack of plant
taxonomy limited his ability to classify and then differentiate among types of
mulberry trees:

> There are two races of Mulberries discerned by these words, blacke and
> white, discordant in wood, leafe and fruite: hauing neuerthelesse that in
> common to spring lalte, the dangers of the coldes being past, and of their
> leaues to nourish the Silk-worme. One sees but one sort of the blacke Mul-
> berries the woode whereof is solid and strong, the leafe large and rude in the
> handling, the fruit black, great, and good to eat. But of the white, there is
> manifestly knowne three species, or softs, distinguished by the onely colour
> of the fruit, which is white, black, and red, so separately brought forth by
> diuers trees, bearing all neuerthelesse the name of white [41, pp. 20-21].

Military Science

Writers of military science books assumed that readers understood the basics of
firearms, defense, and military procedures—most likely passed down from father
to son in many English households. These books provided advanced instruction
in operating large firearms such as cannon, planning and leading formations of
soldiers into major battles, and executing commands. The books describing formations and cannon were usually large folios that contained extensive drawing
aids to help the reader visualize the placement of troops, horses, and artillery.
Readers were expected to have an understanding of military nomenclature, likely
transmitted by oral instruction to all males. Figure 12 shows a page from Robert
Barrett's *The Theorike and Practike of Moderne Warres* [65].

Greek military science applied to English military situations was also popular.
John Bingham, in *The Art of Embattailing an Army* [66], assumed that readers
could read Latin and Greek. With that assumption in mind, he explained Aelian's
military tactics. Figure 13 allows us to see how Bingham carefully defined
terms—in context and in side notes—and referred to the appropriate source, such
as Xenophon. While readers who were not particularly fluent in Latin could read

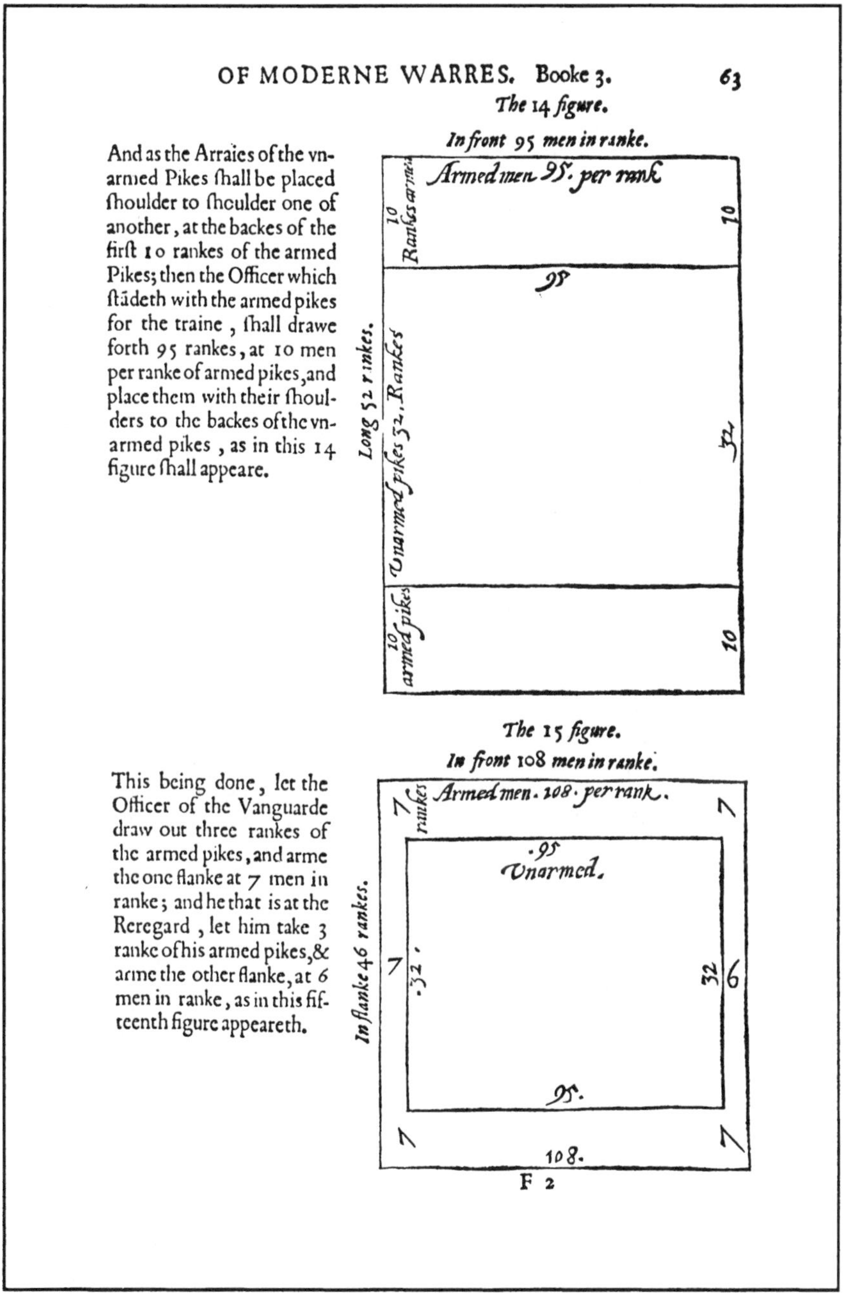

OF MODERNE WARRES. Booke 3.   *63*

The 14 *figure.*

In front 95 *men in ranke.*

And as the Arraies of the vn-
armed Pikes ſhall be placed
ſhoulder to ſhoulder one of
another, at the backes of the
firſt 10 rankes of the armed
Pikes; then the Officer which
ſtãdeth with the armed pikes
for the traine, ſhall drawe
forth 95 rankes, at 10 men
per ranke of armed pikes, and
place them with their ſhoul-
ders to the backes of the vn-
armed pikes, as in this 14
figure ſhall appeare.

Armed men 95. per rank

10 Rankes armed

95

Long 52. rinkes.

Unarmed pikes 52. Rankes

10 armed pikes

10

52

10

The 15 *figure.*

In front 108 *men in ranke.*

This being done, let the
Officer of the Vanguarde
draw out three rankes of
the armed pikes, and arme
the one flanke at 7 men in
ranke; and he that is at the
Reregard, let him take 3
ranke of his armed pikes, &
arme the other flanke, at 6
men in ranke, as in this fifteenth figure appeareth.

Armed men. 108. per rank.

7 rankes

·95
Unarmed.

In flanke 46 rankes.

7   ·32·

32   6

95.

108.

F 2

Figure 12. Page showing an attack strategy, from Robert Barrett,
The Theorike And Practike of Moderne Warres (1598) [65].

the Art of Embattailing Armies. 15

NOTES.

THis Chapter sheweth the inlarging of a *Phalange* or battell ; by diuers placings partly of the armed, partly of the light-armed. It is not hard to be vnderstood ; the rather becaufe moft of the alterations here mentioned are fpoken of heretofore either in *Ælian*; or in my notes : Sixe formes are here fet downe, two by changing the place of fome of the Armed, the other foure by changing the place of the light-armed, the armed are altered by [b] Parembole or [c] Proftaxis, the light armed by [d] Protaxis, [e] Epitaxis, [f] En-taxis, and [g] Hypotaxis ; what the fignification of each is fhall be fhewed in the notes following.

[b] Doubling the front by middle men.

[c] Adioyning.

[d] Forefronting.

[e] Placing after.

[f] Placing betweene.

[g] Placing on the wings.

1. *Parembole.*] This muft alwayes be of armed, which are taken from the reare of the Armed, and inferted betwixt the files of the front : of this kinde is the doubling of the front by middle men with their halfe files, whereof *Ælian* hath fpoken in the 29 Chapter, fee the figure there.

2. *Protaxis* or *fore-fronting.*] I haue fhewed before in the notes vpon the feuenth Chapter, that *the light-armed* were diuerfly placed in the front, in the reare, on the wings, within the battell ; when they are placed before, it is called Protaxis, fee the figure here : [h] *Ptolomie* and *Seleucus* being to fight a-gainft *Demetrius*, who had many Elephants, placed the light armed *before*, to the intent to wound the Elephants and turne them away from their Pha-lange : fo *Alexander*, fo *Darius*, at the battaile of *Iffos*, placed *darters* and *flingers before the fronts of their phalange* : they ferue greatly to annoy the Enemy be-ing fo placed, efpecially being not charged with *horfe* or *pikes* ; if they be charged with either, they are to retire into the interuals of their owne bat-taile of pikes : See *Onofander* cited by me in my notes vpon the 7 Chapter of this Booke.

[h] *Diod.Sic.l.19.* 717.A.

3. *Epitaxis.*] Ordering of the light-armed *behinde* was the vfuall man-ner of the *Macedonian* Embattelling, from whence they drew them at plea-fure to any place of feruice : fee the 7 Chapter.

4. *Proftaxis*] it is, when armed are taken from *behinde* and laid to one or *both flankes* of the battell fronting euen with the front thereof, which is a doubling of rankes, as is before fhewed, done when the hinder halfe files diuide themfelues, march out and front with the fileleaders, or elfe march out entirely without diuifion.

Entaxis] Incifion is alwayes of the light armed into the *fpaces* of the ar-med. It is all one with *parentaxis*, another Greeke word vfed in the fame fence.

Figure 13. Page showing method of defining military terms, from
John Bingham, *The Art of Embattailing an Army* (1629) [66].

the text, the level of the presentation assumed that readers would be interested in understanding and applying a high level of military strategy which had its roots in Greek and Roman military history.

Because these books were aimed at an educated and elite audience, many contained fold-out pages as well as tables explaining trajectories per distance for artillery and numbers of troops for squadrons and battalions that would be sent into battle situations. Military science books were clearly the most technical of books by modern standards. Descriptions are precise; terms are defined; procedures are specific; and visual aids are used extensively to support the meaning of the text.

Smaller reference books designed to be easily portable, such as Achesone's *The Military Garden* [67] and Barnabe Rich's *A Path-Way to Military practise* [68], were available to teach military commanders the correct method of delivering oral directives as well as the responsibilities of different levels of officers. These contained detailed lists of proper commands (see Chapter 3, Figure 30, p. 85) and definitions of terms as well as illustrations to define military formations.

Books on Gardening

As noted earlier, the importance of gardening to English Renaissance readers is nowhere more apparent than in the number and variety of books on this topic— from small pamphlets on kitchen gardens, directed to women readers, to tomes on gardening for vocational gardeners. While the little books on kitchen gardens (perhaps 30 pages) covered information on how to select and plant seeds for selected herbs and vegetables, *The Gardeners Labyrinth* [49] contained 31 chapters that dealt with the full range of tasks plus medicinal applications of plants. Chapter 30 alone, on propagating gourds, is a 2800-word discussion on planting, nurturing, and harvesting these vegetables. For example:

> As the kindes of the Gourdes, requyre the same travaile and diligence in the bestowing in the earthe, as afore uttred of the Cucumbers, which after the large setting asunder, and often watering appeare (for the most parte) aboue the Earthe, by the first or seuenth day after the bestowing in the beds.

> The wrake and teuber braunches, wot up to some heighth, and coueting by a certaine propertie in nature upward, require to be diuersly aided with poles to run up in sundry manners, as either ouer a rounde and vaulted harboure, to giue a more delighte, throughe the shadowe caused by it, and the semely fruites hanging downe, or else by poles directed quite uprighte, in which the Gourde (of all other fruites) moste earnestly desireth, rather than to run braunching and creeping on the grounde like the Cucumber.

> The plants loue a fat, moyst, and dunged loose ground, as the Neapolitan *Rutilius* in his instructions of husbandry hath noted: If a diltgece be bestowed in the often watring of them, the plants require a lesser care and trauaile, in that they are very muche furthered, by the store of moisture, although

there may be found of those, which reasonably prosper with small store of moysture, or being seldome watered, and that they of the same yeelde fruite of a delectabletaste [49, pp. 146-147].

The Latin allusions and the additive, meandering style that preserved the technical detail of the planting process suggest that the work was written for professional gardeners, who were men. The title page, with a woodcut illustration of two male gardeners tending an elaborate garden, reinforces this conclusion. In contrast, Richard Gardiner's *Profitable Instrvctions for . . . Kitchin Gardens* provided more concise, less detailed instruction, as in his description of how to plant seed-bearing parsnips, which uses a conversational style:

Prepare such place in your Garden as is most conuenient for the setting of Parsneps for seeds; first digge and make your ground ready in beds, like as you would sowe any other seedes, then make choice of the fairest Parsnep roote, and plant them in the beds a rowe of rootes on either side the bed, about six inches from the edge of the bed, and a rowe of rootes along the midst of the bed or beds, and set euery roote so neere as you can, to be xv. inches one from another: and when the first seedes doe begin to be ripe, then cut them daily as cause requireth: for the seedes of Parsneps are very apt to fall when they be ripe, to the losse of the best seede (if they be not heedefully looked unto). Thus soone, you shall haue good Parsnep seedes to pleasure any person in that behalfe, otherwise it is not so good nor so profitable [69].

RENAISSANCE TECHNICAL WRITERS AND THEIR READERS

An appreciation of the commercial impetus for Renaissance books and comparisons among books on specific subjects provide students of technical writing with a rich source of information about Renaissance English people.

First, the range and number of extant technical books show us what technologies were important, how work was performed, and how technology advanced during the Renaissance. Changes in books, such as books on silkworm production and on farming and animal husbandry, show that technology and literacy advanced rapidly from 1550 to 1640. The popularity of books such as Tusser's [43], Markham's [44-46], and Fitzherbert's [47] suggests that literacy among readers of agricultural material was still varied but that the demand for books providing more technical detail than Tusser's was increasing.

Second, the popularity of many of these books, indicated by the number of extant multiple editions, suggests that writers were successful in appealing to their audiences. Examining carefully the style, organization, and content of a book can provide some indication of why it may have been successful or unsuccessful. Many large and complex books, such as those written for physicians and for readers within the elite classes who could afford large books, exemplify the range of knowledge available in military science, navigation, and medicine.

Third, the perception that Renaissance technical writers were aware of their readers' content needs and levels of reading comprehension shows that the audience-centered nature of modern technical communication has substantial historical justification. Technical writers' awareness of their readers gives us a means of assessing the literacy of those readers. Because many technical books provided separate sections for men and women and because these showed some differences, we can assume that female readers of the middle classes—those who still had substantial responsibilities for estate management—were able to read non-aureate English.

Fourth, examining a wide range of how-to books and the technical sophistication of books in fields such as navigation, gardening, agriculture, and military science provides a social commentary on what knowledge was valued by English Renaissance people. The large number of books on these topics, as well as on land law, demarcates the concerns of these English readers. In short, technology developed in response to need, and technology and literacy seemed to nurture one another. The emergence of a sophisticated style indicates the need for such a written discourse to textualize advancing information to communicate improvements in technology. Improvements in printing, format, visual presentation, and prose style provide sufficient evidence that English technical books had become a distinctive genre by the closing years of the English Renaissance.

These technical books support a number of observations concerning Renaissance literacy: they were written with specific audiences in mind; they show a distinctive attempt to match content to readers; writers were aware of the educational level of their readers and chose or eliminated Latin and Greek terms accordingly; style for general readers was more unadorned, succinct, and direct than the style of works written for more educated audiences; writers were conscious of the need for proper definition; and writers were conscious of the need for visual aids to enhance their message. As I will discuss in Chapter 6, Renaissance technical writing made good use of advancing technology and changing art techniques in offering improved methods of visualizing information once confined to linear discourse. In short, many modern techniques used to ensure that procedural writing meets the needs of readers were also used by Renaissance English technical writers.

As a preface to a discussion of the evolution of technical style, it is important to note that technical books consistently fell into the category of works using the moderate, low, or utilitarian style. Works of a non-utilitarian nature, such as religious works, used a grandiose style to attain a maximally authoritative presentation. This style combined resources of learning and eloquence into proliferating sentences to produce opulence and weight. In contrast, technical books generally followed the style of the spoken vernacular. Even though the large, elaborate folios written for well-to-do readers who probably had a grammar school and university eduction might contain some elaborately woven clauses and Latinate

words or phrases, the style generally exemplifies intact clausal units and concise presentation.

The interesting point is that technical writers did not see the need for grandiloquence. Technical books emphasized directness in sentence structure, content, and presentation. While appearing stylistically barren, "technical style" was not, in the Renaissance, perceived pejoratively. In the introductions to their books, technical writers did not make excuses for the plainness of their style but instead expressed their hope that the work would meet the needs of readers. Grandiloquence, today admired in Elizabethan literary and philosophical works, was seen by Renaissance writers as appropriate for works written to be politically, philosophically, or spiritually authoritative. Plain style was appropriate for works written to be read and accessed rapidly for instructional purposes.

REFERENCES

1. H. S. Bennett, *English Books & Readers 1475-1557*, Cambridge University Press, London, 1952.
2. L. B. Wright, *Middle-Class Culture in Elizabethan England*, Cornell University Press, Ithaca, New York, 1935.
3. R. Hirsch, *Printing, Selling and Reading, 1499-1550*, Harrasowitz, Wiesbaden, Germany, 1975.
4. H. S. Bennett, *English Books & Readers 1558-1603*, Cambridge University Press, London, 1965.
5. J. Fitzherbert, *Here begynneth a ryght frutefull mater: and hath to name the boke of surueyeng and improumetes*, London, 1523 [STC 11005].
6. F. R. Johnson, Latin versus English: The Sixteenth-Century Debate over Scientific Terminology, *Studies in Philology, 41*, pp. 109-135, 1944.
7. A. Vesalius, *De Humani Corporis Fabrica Libri Septem*, Basilea, ex officina Joannis Oporini, Anno salutis reparatae 1543, Mense Junio.
8. T. Gemini, *Compendioso Totius Anatomie Delaneatio Aere Exarata*, N. Udall (trans.), London, 1553 [STC 11750].
9. R. Record, *The vrinal of physick*, London, 1547 [STC 20816].
10. A. Read, *A Description of the Body of Man*, London, 1634 [STC 20783].
11. A. Read, *The Chirurgicall Lectures of Tumors and Vlcers*, London, 1632 [STC 29781].
12. J. Banister, *A Needefull, new, and necessarie treatise of chyrurgerie, briefly comprehending the generall and particular curation of Vlcers*, London, 1575 [STC 1360].
13. R. Banister, *A Treatise of One Hundreth And Thirteen Diseases Of the Eyes, and Eye-Liddes*, London, 1662 [STC 1362].
14. W. Clowes, *A Profitable and Necessarie Booke of Obseruations, for all those that are burned with the flame of Gun power*, & c., London, 1596 [STC 5442].
15. T. Dawson, *The good huswifes Iewell*, London, 1596 [STC 6392].
16. C. Gesner, *The newe iewell of health*, London, 1576 [STC 11798].
17. P. Slack, Mirrours of Health and Treasures of Poor Men: The Uses of Vernacular Medical Literature of Tudor England, in *Health, Medicine and Mortality in the*

Sixteenth Century, Cambridge University Press, Cambridge, England, pp. 237-274, 1979.

18. T. Elyot, *The Castle of Helth*, London, 1536-1539 [STC 7643].
19. A. Borde, *The Breuiary of Helthe*, London, 1547 [STC 3376].
20. *Here begynneth a newe boke of medecynes intytulyd the treasure of pore men*, R. Bankes, London, 1526 [STC 2199].
21. T. Moulton, *This is the Myrour or Glasse of Helth*, London, 1539 [STC 18214].
22. J. Goeurot, *The regiment of life*, London, 1550 [STC 11970].
23. T. Cogan, *The hauen of health*, London, 1584 [STC 5478].
24. A.T., *A Rich Store-House or treasury for the Diseased*, London, 1596 [STC 23606].
25. E. Roesslin, *The Birth of Mankinde, Otherwyse Named the Womans Booke*, T. Raynald (trans.), London, 1598 [STC 21160].
26. *Here Beginneth the Seynge of Urynes*, W. Powell, London, 1562 [STC 20816].
27. W. S. C. Copeman, *Doctors and Diseases in Tudor Times*, Dawson's of Pall Mall, London, 1960.
28. T. Elyot, *The Boke Named the Gouernour*, London, 1531 [STC 7635].
29. *A very propertreatise, wherein is briefely sett forth the are of Limming*, R. Tottill, London, 1581 [STC 24253].
30. L. Mascall, *A profitable boke declaring dyuers approoued remedies to take out spots and staines*, London, 1583 [STC 17590].
31. W. P[hilip?] (trans.), *A booke of Secrets*, London, 1591 [STC 3355].
32. H. Platt, *The iewell house of art and nature*, London, 1594 [STC 19991].
33. H. Platt, *Delightes for Ladies, to adorne their Persons, Tables, closets and distillatories*, London, 1602 [STC 19978].
34. [Book of Cookery], R. Pynson, London, 1508 [STC 3297].
35. J. Murrel, *A Daily Exercise for Ladies and Gentlewomen*, London, 1617 [STC 18301].
36. J. Partridge, *The widdowes Treasure*, London, 1595 [STC 19434].
37. J. Partridge, *The treasurie of commodious conceites and hidden secrets*, London, 1584 [STC 19426].
38. M. D. Dale, The London Silkwomen of the Fifteenth Century, *Economic History Review, 4*, pp. 324-355, 1932-1935.
39. T.M., *The Silkewormes, and their Flies*, London, 1591 [STC 17994].
40. W. Stallenge, *Instrvctions for the increasing of Mulberie Trees, and the breeding of Silke-wormes, for the making of Silke in this Kingdome*, London [STC 23138].
41. O. DeSerres, *The perfect vse of silk-wormes and their benefit*, London, 1607 [STC 22249].
42. C. C. Camden, *The Elizabethan Woman*, Paul P. Appel, Maroneck, New York, 1975.
43. T. Tusser, *Fiue hundreth points of good husbandry*, London, 1597 [STC 24385].
44. G. Markham, *Countrey Contentments*, London, 1615 [STC 17342].
45. G. Markham, *Cheape and Good Hvsbandry*, London, 1614 [STC 17336].
46. G. Markham, *The English Arcadia*, London, 1606 [STC 17351].
47. J. Fitzherbert, *Booke of Husbandrie*, London, 1598 [STC 11004].
48. W. Lawson, *A New Orchard and Garden*, London, 1623 [STC 15329].
49. T. Hill, *The Gardeners Labyrinth*, London, 1577 [STC 13485].

50. I. Fletcher, *The Differences Cavses and Ivdgements of Vrine, According to the Best Writers Thereof, Both Old and New, Summarily Collected*, London, 1622 [STC 11063].

51. A. Paré, *An Explanation of the Fashion and Vse of Three and Fifty Instruments of Chirvrgery*, London, 1634 [STC 19190].

52. G. Abbott, *A Brief Description of the Whole Worlde*, London, 1599 [STC 24].

53. *Mappa Mundi*, R. Wyer, London, 1535? [STC 17197].

54. I. Turler, *The Traveiler*, London, 1575 [STC 24336].

55. R. Eden, *The Arte of Nauigation*, London, 1589 [STC 5798].

56. W. Bourne, *A regiment for the sea*, London, 1580 [STC 3425].

57. P. Garcia, *The Rutter of the Sea*, London, 1555 [STC 11552].

58. L. Wagner, *The Maringers mirrour*, A. Ashley (trans.), London, 1588 [STC 24931].

59. R. Norman (trans.), *The Safegard of Sailers*, London, 1584 [STC 1581].

60. W. Johnson, *The Light of Navigation*, London, 1612 [STC 3110].

61. W. Cunningham, *The Cosmographical Glasse, Conteinyng the Pleasant Principles of Cosmographie, Geographie, Hydrographie, or Navigation*, London, 1599 [STC 6119].

62. L. Mascall, *The government of Cattell*, London, 1614 [STC 17586].

63. L. Mascall, *The husbandlye ordring of poultrie*, London, 1581 [STC 17589].

64. L. Mascall, *A booke of the arte and maner, how to plant and graffe all sortes of trees*, London, 1572 [STC 17574].

65. R. Barrett, *The Theorike and Practike of Moderne Warres, Discourses in Dialogue wise*, London, 1598 [STC 1500].

66. J. Bingham, *The Art of Embattailing an Army*, London, 1629 [STC 162].

67. J. Achesone, *The Military Garden, Or Instrvctions For All Young Sovldiers*, London, 1629 [STC 88].

68. B. Rich, *A Path-Way to Military practise*, London, 1587 [STC 20995].

69. R. Gardiner, *Profitable Instrvctions for the Manvring, Sowing and Planting of Kitchin Gardens*, London, 1603 [STC 11571].

CHAPTER 5

English Renaissance Technical Writing and the Emergence of Plain Style: Toward a New Theory of the Development of Modern English Prose

Throughout the last two chapters, I have discussed the issue of style as it occurred in Renaissance technical writing. As the excerpts included in Chapter 4 show, technical writers used technical nomenclature, as it existed and was appropriate to the comprehension level of their readers. In many works, such as herbals, description was limited to precise, highly visual adjectives and common nouns because botany as a discipline had yet to emerge. While first-aid books, for example, used common terms reminiscent of daily speech, medical books for practitioners used a combination of Latin and Greek nomenclature and vernacular terms. Medical books, such as those written by Alexander Read, clearly show that physician writers educated on the European continent saw the need to teach Latin medical terminology to "un-Latined" English physicians and often used a Latin or Greek term followed by a vernacular definition or synonym. At this point, an examination of the syntax of English Renaissance technical writing is in order.

RENAISSANCE TECHNICAL WRITING:
A GENRE IGNORED

Studies in the development of modern English prose style have addressed the rise of Ciceronian, Senecan, and Attic styles; but these studies have developed their conclusions by analyzing only major literary forms: character and biography, fiction, drama, sermon, essay, and history as these genres reflect various uses of Ciceronian and anti-Ciceronian styles. Similarly, these studies focus on traditional authors such as Glanville, Hooker, Bacon, Montaigne, Erasmus, Chaucer, More, Ascham, Sydney, Peacock, and Caxton in tracing the shifts from a florid "Elizabethan" style to the "plain style" advocated by the scientists of the Post-Restoration Royal Society. In their landmark assessments of the rise of English style, both Morris Croll [1] and R. F. Jones [2] presented their theories of modern prose style by focusing only on works recognized, prior to 1950, as fitting within the traditional literary canon.

Croll argued that anti-Ciceronianism became the dominant prose style of the seventeenth century, "the rhetorical and literary expression of science" [1, p. 195]. Croll stated flatly that Montaigne and Bacon were

> the first writers in the vernacular languages who employ a style which renders the process of thought and portrays the picturesque actuality of life with equal effect and constantly relates the one to the other. . . . Bacon, Hall, Jonson, and Wotton in England—are the actual founders of modern prose style [1, p. 184].

Croll then asserted that

> In the works of these authors, and in none of those that precede them, we can find a style in the popular language which is at once firm, uniform, and level enough to be called a style and also adaptable enough to adjust itself to the changing life of the modern world—a style which may grow and change in later generations without losing its recognizable features [1, p. 184].

Jones, in contrast, argued that the shift occurred about 1660 and resulted from the influence of the new science and its Baconian aspects on language [2]. In *The Triumph of the English Language*, Jones discussed, from the perspective of Renaissance literati, the problems stemming from their belief that vernacular English, as opposed to Latin, was inadequate and ineloquent [3]. Jones argued that the introduction of inkhorn terms was considered by many of the intelligentsia as necessary to improve the capacity of English to communicate fully the multi-textured meaning of Latin works. In discussing English as a useful, practical language, Jones discussed only the growth of educational books, such as rhetorics, translated from Latin. He defined "popular instruction" as works dealing with religious and humanistic learning [3, pp. 32-67].

This narrow emphasis on describing the emergence of modern English prose by referring only to genres accepted in literary studies, I will argue, has three deficiencies: it ignores a large segment of published writing in the English

Renaissance; it obscures the long tradition of English plain style that existed before 1600; and it ignores the contribution of technical books to the emergence of utilitarian discourse as defined by Bacon and later by writers such as Spratt and Glanville.

Subsequent studies precipitated by these two works [1, 2], clearly important in studies of English literary style, have continued the tradition of focusing style analysis only on canonical works in literary genres. Subsequent authors have thus described development of English style from within either the Jones camp or the Croll camp [4-9]. More recently, both A. D. Hall [10] and Adolph [11], for example, have expanded Croll's and Jones's positions, but they have continued to emphasize classical rhetoric as it influenced accepted major Renaissance writers. Fish [12], in contrast to Williamson [13], a Jones disciple, has offered a different description of the rise of modern English style; but even Fish, a deconstruction advocate, has limited his analysis to writers included in the English canon. The efforts of these representative scholars to track changes in style, from a medium that calls attention to itself to a medium that becomes a vehicle of pragmatic communication, share at least one common perspective: English style was influenced predominantly by classical philosophy, classical rhetoric, and ultimately the philosophy of science. By examining only those works included in the traditional English canon, these scholars, like Croll and Jones, have ignored the substantial numbers of practical works that infused the lives of English people and the possible influence of these texts on the evolution of English style as it actually existed in a variety of written texts.

Adolph [11] in 1968, Schlauch [14] in 1959, Gordon [15] in 1966, and Mueller [16] in 1984—to name four fairly recent works attempting to take a different approach—have sought to expand studies of the development of modern prose style by recognizing the broad body of utilitarian style and utilitarian works usually by-passed by historical studies of English style. Adolph believed that science has been given too much credit for style changes in the late seventeenth century because utilitarian writing provided a powerful presence [11, pp. 162-163]. Similarly, Schlauch [14] and Gordon [15] argue for the inclusion of chronicles, wills, charters, guild records, familiar letters, and even recipes and written instructions. (Many of these, as indicated in earlier chapters, fall clearly within the province of "technical" writing.) Mueller agreed with their perspective, describing the use of modern English style in Renaissance self-help medical books [16, pp. 281-287], which she categorized as the "prose of counsel." Mueller's thesis affirmed the value of studying prose as "a self-contained body of materials" [16, p. 5]. However, Mueller argued that Scripturalism, defined as "a writer's absorption with the text of the Bible and with rendering its meaning in English—an absorption so intense as to mark the writer's own style with the impress of Biblical modes of expression," was the driving force for the use of modern English in an increasing number of written documents [16, p. 40].

While the legacies of deconstruction and postmodern critical theories have invited students of language to examine the merits of works outside the canon, traditional scholars such as Chambers (1932) [17] and Mueller (1984) [16] both recognized that a direct English style can be found in numerous early English writers—such as Hilton, Rolle, and Wycliffe—long before Renaissance philosophical discussions surfaced to debate the value of English as a suitable vehicle for learned discourse. My point is, therefore, that substantial critical studies and textual artifacts exist to justify further questioning the view that Bacon's work is the basic point of departure for understanding the development of English prose style.

In addition, the quantity and variety of non-canonical books, as revealed by even a casual perusal of the *Short-Title Catalogue*, illustrate the questionable nature of traditional stylistic studies, resting on Croll and Jones, that assume that intellectual history based on selected canonical texts is the prime determinant of literary history, which in turn is the ultimate measure of what modern language students should believe about the roots of our language. As Bennett [18, 19], Wright [20], and Heninger [21] have convincingly shown, English readers were enjoying an ever expanding assortment of books, many of them written by educated writers who either 1) believed that books on a variety of topics should be made available to English readers or 2) saw the financial reward of providing books that would appeal to a newly literate audience eager for works in the vernacular. While medieval and Middle English scholars have had to deduce language development from surviving incunabula primarily written and preserved by religious or government institutions, students of Renaissance prose have the benefit of a large body of extant published writing that emanated from outside the confines of government, the church, and the academy. To ignore the contribution of this writing to the development of modern prose is to try to assume that only learned discourse influenced English society, a conclusion that is certainly not warranted by English history.

RENAISSANCE TECHNICAL WRITING AND ITS RELATIONSHIP TO MODERN ENGLISH PROSE DEVELOPMENT

In arguing for an expanded approach to the development of what I will call English plain style—an approach that needs to include a much larger collection of books than those recognized by literary historians—I wish to make six observations which will be the focus of this chapter:

1. As discussed in Chapters 2 and 4, many English Renaissance technical writers also wrote literary, religious, and philosophical works. Thus, these writers, in their choice of style for their technical books, were aware that they were writing in a different genre. As we have already seen in Chapters 2 and 3, these early technical writers chose the method of presentation—format, content, and page design—that they believed best suited the genre and reader for whom they were

writing. In a sense, the functional style of Renaissance technical books was an integral part of the functional aspects of format and page design as they reflected the basic reading needs of English readers.

2. The use by many of these writers of a style that would eventually be attributed to Bacon definitely emerged before Bacon's works on language became a force during the later years of the seventeenth century.

3. Contrary to the belief that science was the major impetus behind the shift to utilitarian prose style by the latter years of the seventeenth century, English Renaissance technical books were already using what Chambers called "Alfredic English" [17], or what Adolph labeled the utilitarian style derived from Socrates' *genus humile* [11], or what Mueller called the "medium" modern style [16]. These first writers of English technical books were adopting a style devoid of the self-conscious use of rhetorical devices. They used the vernacular to convey information clearly and directly rather than to heighten the affective qualities of the prose.

4. Bacon's beliefs—that the audience ultimately governs communication, that the kinds of discourse have their own ends, that form and structure must be functional—can be seen operating in technical manuals written throughout the sixteenth century. In short, Bacon was formally articulating as theory what Renaissance technical writers and Biblical writers, such as Wycliffe and Tyndale, had already been doing. Like Wycliffe, Tyndale, and Bacon, these early English technical writers were not interested in words for their own sake but for the ways in which they could be used to present fact and truth to English readers.

5. Renaissance English technical writers, long before Bacon, saw the value of what would come to be called the plain style (or *genus humile*) revered by the Royal Society—a style in which the writer himself intruded as little as possible. Yet, they also employed a variety of styles that, when appropriate, revealed the attitudes of the writer. As discussed in Chapter 4, in books written for two distinct audiences, long before the close of the Renaissance, we can see different styles in the same technical manuals—further indication that these writers believed that different segments of technical books required different styles.

6. The distinctiveness of technical books as they anticipate major stylistic shifts in the later Renaissance and then in the seventeenth century suggests that technical writing can be identified as a fifth genre operating in the Renaissance: comedy, satire, epigram, epistle, and technical manual. As Wesley Trimpi has shown, the intention of the original four "was to reveal with the greatest possible candidness and accuracy what men actually do" [22, p. 127]. Technical writing has a concomitant purpose in its process analysis and procedure writing, both of which exist to help readers perform tasks.

To examine the rationale for each of these assertions requires 1) that we find a suitable definition for modern plain style, 2) that we examine support for the existence of a plain English style in medieval English documents, 3) that we track its existence in a broad selection of English technical books of the fifteenth,

sixteenth, and seventeenth centuries, and 4) that we note the similarity of this style as it parallels the goals of prose defined by Bacon and then the Royal Society. In suggesting the role of Renaissance technical writing in the development of modern English prose, I will build on work by writers such as Mueller [16] and Adolph [11], who have stated their awareness that the utilitarian tradition was a force in English prose. However, in expanding on their work, I will show how the style of technical writing adds an important ingredient to current knowledge of modern English prose development.

PLAIN ENGLISH PROSE—A DEFINITION

The term "plain" English, from a Renaissance perspective, denotes a written English prose characterized by precise word selection and an attempt by the writer to provide ease of comprehension for the reader, who was likely able to read only vernacular English. Renaissance plain style can also be characterized by its common, colloquial vocabulary and lack of purposeful sound effects resulting from tropes and deliberately sententious schemes. The aim of plain English was to be denotative, to enable both writer and reader to share the same meaning conveyed by the sentences.

Because "meaning" in English rests less on the word level than on the clause or period level, the "plain" English clause employs what Mueller, quoting Bever, termed "canonical sentoids—an intact clause that emerges in the s-v-o [noun phrase + verb + noun phrase] that also reflects spoken English speech patterns" [16, p. 5]. Historical linguists have found that the v-o order was dominant by 1300 and firmly fixed for both independent and dependent clauses in prose texts by 1400. Declarative (s-o) order and interrogative (v-s) order became distinct types during the same period [16, pp. 5-6]. As McLaughlin noted:

> For both transitive and intransitive sentences the order subject-verb clearly predominates in the late fourteenth century, as does the order subject-verb-object. Apparently, too, at this stage in the development of word-order patterns, such order was not, as it was to some extent in Old English, contingent upon whether a given sentence is or is not embedded in another [23, pp. 242-243].

As Schlauch [14] and Earle [24] have also argued, modern prose style incorporated spoken English into its word order, as s-v-o also characterizes speech. Written and spoken English merged after the Norman Conquest because of the similarity of the spoken and written forms. Bolton [25, pp. 101-102] and other linguists have concluded that English syntax, rather than Latin syntax, sustained its power in written English and became the standard syntax for all kinds of documents by the late seventeenth century because Latin was never spoken in England to any extent. Again, the spoken language became the dominant language: because language reflects thought, written Latin was never as congenial to

writers, who translated Latin into the vernacular, particularly when the audiences for many of their works were literate only in native English.

Besides its s-v-o word order, modern English syntax emerged long before the Renaissance and exhibited a number of characteristics now documented by linguists:

- speech-based connectives: and, for, so, but, then;
- cumulative patterns resulting from connected sentoids;
- asymmetric rather than symmetric sentence units built with connectives;
- common, colloquial diction;
- lack of sound-affected words and phrases;
- decreasing use of doublings, except for clarity;
- avoidance of loaded sentence patterns;
- emphasis on the clause/sentence rather than the Latin period;
- metaphors and examples for clarity rather than for sensuous affectiveness.

Based on this amalgamation of definitions of modern prose (which literary scholars believe did not exist in Renaissance England, and has surfaced only through the efforts of twentieth-century linguists, philologists, and rhetoricians), I now want to examine modern prose syntax as it occurred in pre-Renaissance writings, then track its development in a number of representative English Renaissance technical books. Ultimately, I will show how these characteristics of modern English prose style, already fully operative in instrumental discourse by Bacon's time, were absorbed by Bacon into his own theory of communication. The communicative value of this modern prose style, described by Bacon and then advocated and refined by the Royal Society, was simply an extension of evolving style characteristics that emerged throughout the English Renaissance. Bacon has received credit for this style because he was the first to articulate its characteristics into a workable theory of language. Bacon supported this language because it aided his efforts to advance science.

PRE-RENAISSANCE ENGLISH AND THE
PERSISTENCE OF PLAIN STYLE

To establish a basic framework for examining English plain style as it emerged in pragmatic discourse of the English Renaissance, it will be helpful to track the continuity of plain style in Old English and Middle English writing and then in translations of the Bible and Caxton's early books. While such a discussion can only highlight significant works, it can illuminate a major point: what we now call clear or plain style, the focus of extensive style research, was part and parcel of the continuum of English prose. It exemplifies the type of style that was non-aureate

and was used for writing that was functional in a variety of pragmatic and non-pragmatic contexts.

Old English Plain Style

In his classic study of the continuity of English prose from Alfred to More, Chambers was one of the first language historians to argue that "Alfredic" English, as it emerged in the tenth century, persevered despite periodic attempts to augment English for the purposes of eloquence [17]. Chambers compared Alfred's direct style with an Elizabethan translation of the same passage. Alfred wrote, "Chiefly he went thither (as well as for the surveying of the land) for the horse-whales [walrus], because they have very fine bones in their teeth." The Elizabethan translation in Hakluyt's *Navigations* phrases the same passage as follows:

> The principal purpose of his travel this way, besides the increase of the knowledge and discovery of those coasts and countries, was for the more commodity of fishing of horse-whales which have in their teeth bones of great price and excellence [quoted in 17, p. lx].

Chambers's point here was that the clotted style of Elizabethan prose was essentially alien to what would come to be called seventeenth-century plain style. He supported this argument by showing that the Alfredic tradition continued to surface in documents—for example, in the *Anglo-Saxon Chronicle* in the eleventh century, in Hilton in the fourteenth century, and again in Berners in the fifteenth century. While Chambers's work has been refined, his point—the persistence of English plain style, its curt syntax and speech-based character—has been widely accepted by critics willing to examine English prose development from a wider perspective than canonical texts. Mueller finds moderate style in these works and in others such as in Nicholas Love's *Myrrour of the Blessed Lyf of Jesu Christ* and in Old English Biblical renderings [16, pp. 56-57]. I argue that this moderate style has what technical communication students would now associate with qualities of clear, modern technical style. Mueller saw the value of this style in religious works for a broad audience. But I would add that this audience included the newly literate reader who needed, from a comprehension perspective, a direct, unaffected, idiomatic English. This reader was more interested in understanding the words than in experiencing mystical transport [16, p. xx]. However, this same style also appealed to a broad audience needing written instructions and descriptions to be able to perform processes.

Chambers then alluded to the syntactical strength of the eleventh-century *Anglo-Saxon Chronicle* account of the quarrel between Earl Godwin and Edward the Confessor, which illustrates short interlocking clauses:

> They [the Northern lords] were so much of one mind with the kind, that they would have attacked the army of Godwine if the kind had wished it. But

some men thought that it would be great folly, were the two sides to come together in battle; well nigh all that was most gallant in England was in one or other of the hosts. Such men thought that they would be making an open way to oust enemies into the land, and be bringing to pass great destruction among ourselves [quoted in 17, p. lxiii].

Middle English Plain Style

Chambers found this same forceful style again surfacing in the twelfth century in religious prose, such as *The Ancren Riwle*, which uses character drawings that anticipate those that would appear in sixteenth-century farming books, such as Fitzherbert's *Booke of Husbandrie* (discussed later in this chapter), as well as character essays. In *The Ancren Riwle*, the backbiter is described thus:

> He casts down his head, and begins to sigh before he says a word; then he talks around the subject for a long time with a sorrowful countenance, to be the better believed: "Alas, well-away, woe is me, that he (or she) has fallen into such repute." Enough did I try, but I could do no good herein. It is long ago that I knew of it; but nevertheless it should never have been betrayed by me; but now that it is so widely known through others, I cannot gainsay it. They say that it is bad; and yet it is worse than they say. Grieved and sorry I am that I must say it; but in truth it is so, and that is a great grief. For many other things he (or she) is greatly to be praised; but nor for these, and woe is me therefore. No one can defend them [quoted in 17, pp. xcvi-xcvii].

A similar style—characterized by short, active-voice clauses, colloquial diction, and cumulative sentences formed by sentential conjuncts—would become the standard of late seventeenth-century prose. For example, in the 1676 revision of Glanville's *The Vanity of Dogmatizing*:

> But our Affections misguide us by the respect we have to others, as well as by that we bear to our selves: I mentioned that Instances of Antiquity, and Authority. We look with superstitious Reverence upon the accounts of past Ages, and with a supercilious Severity on the more deserving products of our own: a vanity that hath possest all times as well as ours, and the golden Age was never present [quoted in 11, p. 86].

Moving backward in time some three hundred years, this same chisled prose can also be seen in the work of Richard Rolle of Hampole, whom Chambers described as one of the most popular English writers in the late fourteenth and the entire fifteenth century. Rolle's style shows clear affinities with the *Anglo-Saxon Chronicle* and *The Ancrene Riwle*:

> B. The Bee and the Stork.

> The bee has thre kyndis. One is that she is neuer idle, and so is not with them that will not work, but casts them out, and puttes them awaye. Anothers is that when she flys she takes erthe in her feet, that she be not lightly

overhead in the air of wind. The third is that she keeps clene and bright her wings.

Thus rightwise men that lufes God are never in ydlenes; For either they are in travail, praying, or thanking, or reading, or other good work; or withtake idle men, and shew them worthy to be put from the rest of heven, for they will not travel here [quoted in 26, p. 231].

Chambers and numerous later language historians have noted that English became the medium of expression by the middle decades of the fourteenth century. In 1362 the law courts were ordered to use English for their proceedings; in 1363 and 1365 Parliament was opened in English; and in 1399 Henry IV accepted the crown with a speech in English. A southwestern deed of 1376 is the oldest private legal instrument surviving in English; the oldest petition to the Crown is that of the mercers' guild of London in 1386; the oldest English wills in the London Court of probate date from 1387; and in 1389, returns of the ordinances, usages, and holdings of the guilds were made in English [16, p. 9]. As Chambers also noted, English used a serried, terse form when it emerged in the fifteenth century; after 1420, it became a popular vehicle for writing about common affairs, such as complaints about unclean roads:

Also that a mud wall in the bailly by the hie strete, bytwene the house of Shelhard habirdassher and hay Sporyer, fallith doun gobet-mele into the hie strete, and makith the wey foule, in desese of al folk ther passyng and dwellyng [quoted in 17, p. aaa].

Fifteenth-Century Plain Style

It is useful to note at this point that descriptions of the pervasiveness of fifteenth-century English have been further expanded by John Fisher [27] and Malcolm Richardson [28], who have argued that English emerged in this century as an administrative language defined by the Chancery clerks. As Fisher stated, "The Rolls of Parliament were entirely in the hands of the Chancery clerks, and nothing they wrote could be more official or prestigious." As a result, "this variety of fairly modern, fairly standard prose, was in wide use in the Westminster offices by the 1430's" [27, pp. 880-881]. An example petition, cited by Fisher, shows its affinity with the direct, colloquial style of the previous examples. Note the lack of punctuation to separate sentences and the influence of French legal phrasing that mingles with the use of connectives to build periods:

[Heading] To the kyng oure soueraigne lord [Text] Besechith mekely leuin le Clerc Burgeis of Gand to consider by encheson of the trewe acquitaille that the seid towne hath doon and doth dayly in diuerse maners vnto yow oure souereign lord howe that but late ago he boght of Robert Brampton of Caleys attorney to his brother William Brampton of Chestreville in Derby shire certain wolles to the valueof xijc and xv nobles the whiche monoie the same leuin paied in hand to the seid Robert noght havyng liuerey of the seid wolles

> But the seid Robert bonde his seid Brother and hym selfe in the abouve said somme to the same leuin to the entent that at a certain day he shuld have had lyuerey of the wolles aboue said Notwythstondyng the whiche bonde the seid leuin hath not as yet hadd nother lyuerey of the seid wolles ne of the seid money [quoted in 27, pp. 881-882].

Richardson, extrapolating on the work by Fisher, then showed that Chancery English, because of its influence, also defined the style of business and private correspondence. The art of letter writing was manifested as the *ars notaria*, the science of drawing up legal, business, and diplomatic documents: "the dictamin predictably formed the basis of virtually all the documents the chancery issued—royal writs, indentures, treatises, diplomatic correspondence, charters, grants, pardons, oaths, inquisitions, regulations, summons to Parliament, and innumerable other forms of parliamentary proceedings" [28, p. 210]. In short, what would ultimately be called plain style in seventeenth-century England was entrenched in legal writing by the middle years of the fifteenth century.

Richardson traced the similarity in style by examining a Chancery warrant of the reign of Henry V (1417) and examples from the Plumpton, Paston, Cely and Stoner letters (1479):

> Worshipful fader in God, right trusty and welbeloued, We grete yow ofte tymes wel. And for as muche as we haue vnderstande that Maister Iohan Chaundeler that was Deene of Salesbury is chosen Bisshop of the same chirche: wherof we hold vs we agreed, and therto we yeue our assent Roial. And we wol wel that after the consecracion of the said Elit, we haue liueree of his temporaltees. And the Holy Goost haue yow in his keping. Yeuen vnder our signet in our hoost afor Faloise the vii day of Decembre [Chancery warrant (1417), quoted in 28, p. 215].

> I grete you wyll. I lete you wyte I haue resayuyd of John Forner and Hary Demorys for the full payment of Phelepe Seller ys letter of payment, wherefor I wyll that ye dellyuer to the sayd John or Hary, the bryngar of thys bill, the plege of Harys the weche Phelype Seller lefete wyt you at Caleys. Wryt at London the x day of Desembor [Cely letters (1479), quoted in 28, p. 216].

Both Fisher and Richardson approached, in the late 1970s, what seems to me to be the issue ignored in existing studies of English style: that modern English developed outside and despite the English literary tradition that focused on Latin. As Fisher concluded,

> Chancery English grew up outside the orbit of either the church or the schools, nearly all of whose reading and writing continued to be in Latin until well into the sixteenth century. In the absence of any other national model for writing in the vernacular, and in view of the enormous prestige and ubiquitous presence of Chancery writing, it is not surprising that chancery set the fashion for business and private correspondence [27, p. 891].

As I stated in Chapter 1, the emergence of England as a commercial power existed alongside the emergence of the English literary Renaissance, marked by the desire of philosophical and literary writers to augment English, thereby making it more capable of an eloquence associated with Latin and French. Thus, by the beginning of the English Renaissance, two distinct styles emerged (with variations on each): one suited for dealing with utilitarian matters important to a populace literate only in English; the other, a highly affected, pretentious, and inefficient style which was often found in the writings of the English literati.

Ultimately, the growth in numbers, activity, and power of a manufacturing, commercial, and bureaucratic section of the population and the organization of guilds, trading companies, municipal corporations, and a civil service during the fourteenth and fifteenth centuries led to the entrenchment of an English-speaking political and economic power base which was formerly controlled by a nobility and the ecclesiastical hierarchy. Modern trends embracing urbanization, secularization, and increasing popular participation in the national economy and culture became motives for various kinds of English prose texts in this period. By the sixteenth century, the inability of the common English person to speak or read either Latin or French, the improving literacy of the middle class, the increasing availability of cheaply printed books for a newly literate population eager for works in the vernacular, and the growth of knowledge in nearly every field—all explain the increasing prevalence of vernacular English as the prose of utility.

The relationship between English as an actively written and spoken medium and the character of the language as it emerged has also been supported by MacDonald [29] and Bennett [30], who recognized the existence of plain style before 1600 and the increase of plain style between 1600 and 1660. Bennett noted that "a good deal of our English has corresponded to our spoken language directness and lucidity" [30, p. 283]. A study of selected technical documents, many of them introduced in previous chapters, suggests two points: 1) plain English, as it was practiced in written discourse, predated Bacon's views on language; but 2) in the seventeenth century, that practice was finally accompanied by a clearly articulated theory of plain style as it was brought to bear on the search for knowledge.

Biblical Plain Style

Before pursuing the development of plain, or moderate, style in Renaissance technical books, we need to examine the contribution of Wycliffe and Tyndale to this tradition because of the long shadows both cast in the rise of vernacular translations and because both reveal a commitment to the speech-based prose already established in Old English.

Wycliffe's translations were composed of cumulative, asymmetrically progressing sentences, which language scholars have shown to be characteristic of modern English prose. To show that Christ's message is infinitely reasonable and

therefore worthy of belief, Wycliffe used structure and syntax to emphasize the logic of the Word. He frequently used enumeration in his sermons to unfold interpretation so as to reveal the logical connections in its meaning. For example, in unfolding the meaning of I Corinthians 13, Wycliffe used enumeration to dissect the chapter into "sixtene condiciouns by which men may knowe . . . charite":

> The thirttenthe condicioun of this love is that it trowith all thingis; ffor things and truethe is al oone, and so al treuthis ben trowid of it. . . . The fourtenthe condicioun of this love is that, it hopith alle thingis; for it hopith that ordeyned treuthe helpith to alle good men, and this charite hopith to have parte of this helpe. . . . The fiftenthe condicioun of this love is that, it susteyneth al thingis; for it helpith to holde al treuth, and abidith the ende therof. . . . The sixtenthe condicioun and the laste that folwith this charite, is that it fallith never awey, neither in this world ne in the tother [quoted in 16, p. 51].

Another way in which Wycliffe emphasized the logic of Christianity was through his use of polysyndeton—repetition of connectives. The synthesizing and combinatory power of sentential conjunction enforces the cohesive logic of scriptural truth. Wycliffe's sermons, with their emphasis on syntax as the unifying method, illustrate the avoidance of rhetorical excess and the locution forms of actual speech.

In the study of modern prose style, Tyndale's Biblical translation is important because it became embedded in existing English translations, and his style exemplifies the colloquial use of speech and cumulative sentences built with conjuncts. That Tyndale's style reflects but then moves beyond Wycliffe's can be seen in a comparison of their translations of Matthew 17:9-11:

> 9   And, hem cummynge doun fro the mounteyne, Jhesus comaundide him, seyinge, Saie ye to no man the visioun, til mannes sone ryse again fro dead.
> 10   And his disciplis axiden hym, seyinge, What therfore seyn scribis, that it behoueth Hely first come?
> 11   And he answerynge seith to hem, Forsothe Hely is to come, and he shal restore alle thingis [Wycliffe Bible (ca. 1384), quoted in 25, p. 137].

> 9   And as they cam doune from the mountayne, he charged them, sayinge, Se that ye shewe thys vysion to no man, tyll the sonne of man be rysen ageyne from deeth.
> 10   And hys disciples axed off him, sayinge, Why then saye the scribes, that Helias muste fyrst come?
> 11   Jesus answered and sayd vnto them, Helias shall fyrst come, and restore all thynges [Tyndale Bible (1526), quoted in 25, p. 186].

That both Tyndale and Wycliffe provide benchmarks in the development of the plain English tradition I have been describing can be seen if we compare Tyndale's exposition of the key verse of Matthew 5, 6, and 7 with Erasmus's translation:

> Seke the kingdome of heaven therfore and the righeousnesse of the same, and
> be sure thou shalt ever have sufficient, and these thynges shalbe ministered
> unto the [Tyndale, quoted in 16, p. 184].

> Lwet your chefe care be aboute that good thynge, incomparison of the
> whiche, these worldly thynges be of no value nor reputacion. The kyngdom of
> god must be set up, that is to say, the doctrine of the gospell, by the which we
> attayne unto the heavenly inheritaunce [Erasmus, quoted in 16, p. 184].

Erasmus's grandiloquence stands in stark contrast to Tyndale's unadorned
statement. Erasmus's style is aureate and uses circumlocutions and loaded patterns. Tyndale, unlike the learned Erasmus, showed no inclination to render
Scripture in anything like the authoritative, aureate style in vogue in Erasmus's
time.

That syntax is crucial to our understanding of the development of modern prose
can be further seen when we examine the passage from Matthew 17:9-11, phrased
into literal modern English, from the Old English Bible. The s-v-o pattern is not
yet fully emergent, but Erasmus is definitely working in that direction:

> 9. And when they from the mountain went, the Savior them commanded, and
> thus said, To none man not say ye this, ere man's Son from death arise.
> 10. And then asked his learning-knights him, What say the scribes that
> it-is-necessary first to come Elias? 11. Then answered he them, Truly Elias
> is coming, and he will renew all things [quoted in 25, p. 96].

Caxton's Styles

After Wycliffe and Tyndale, Caxton is one of the most interesting, if not pivotal,
of stylists because he reflects the ability to use both an aureate style and an
unadorned direct English plain style. Caxton, an English mercer, sought to augment English to make it more permanent by separating it from the influences of
speech. His efforts to eradicate the functionality of language and emphasize an
English that was aureate for its own sake drove his affected use of doublings. The
following excerpt was written by Caxton and added to the prologue to the *Mirror
of the World* (1481):

> Consideryng that wordes ben perishyng vayne and forgeteful/ and writynges
> dwells & abide permanent as I rede Vox audit perit littera scripta manet/ thise
> thinges have ben causes that the faites and dedes of auncyent menn ben sette
> by declaracion in fair and aourned volumes/ to thende that science and artes
> lerned and founden of thionges passed myght be had in perpetual memorye
> and remembraunce [quoted in 16, p. 160].

When Caxton wrote in his own voice, however, he lapsed into the functional
style of speech-based English. In his 1490 translation of the French poem
Eneydos, Caxton stated in his prologue why he decided to examine the original
French version of this poem:

And when I had aduysed me in this sayd boke. I delybered and concluded to translate it in to englysshe and Forthwyth toke a penne & ynke and wrote a leef of tweyne / whyche I ouersawe agayn to correct it / And whan I sawe the fayr & straunge termes therin / I doubted that it shoulde not please some gentylmen which late blamed me sayeng that in my translacyons I had ouer curyous termes which coude not be vnderstande of comyn peple / and desired me to vse olde and homely termes in my translacyons. and fayn wolde I satysfye euery man / and so to doo toke an olde boke and redde therin / and certaynly the englysshe was so rude and brood that I coude not wele vnderstande it. And also my lorde abbot of westmynster ded do shewe to me late certayn euydences wryton in olde englysshe for to reduce it in to our englysshe now vsie / And certaynly it was wreton in suche wyse that it was more lyke to dutche than englysshe I coude not reduce ne brynge it to be vnderstonden / And certaynly our langage now vsed varyeth ferre from that. whiche was vsed and spoken whan I was borne [quoted in 25, p. 173].

Even in lapsing into his native prose, Caxton described his difficulty in translating from French into English and choosing the best style for the translation. Caxton's difficulties reflect the dilemma facing translators who were confronted by those who were committed to Latin's stylistic traditions and others who were committed to translations that used the vernacular and avoided Latinate words.

PLAIN STYLE AND RENAISSANCE
TECHNICAL BOOKS

That this plain style persisted and evolved long before Bacon can be seen in medical books, in herbals, and in books on farming and animal husbandry, and in a variety of books on Renaissance technologies. The prevalence of plain style for various audiences, both learned and unlearned, supports the importance of non-aureate instrumental discourse in understanding the English Renaissance.

Renaissance English Medical Books

The self-help medical books that were so popular throughout the Renaissance offer perhaps the clearest testimony to the popularity and enduring power of English plain style. As Janel Mueller noted, using the vernacular to bring to the "non-Latined" readers ways of improving daily life was as important to Renaissance English writers as providing readers with vernacular works on religion and spiritual living [16, p. 119]. The constant threat of disease from plague, food, wounds, and assorted afflictions whose causes were unknown but feared was likely the cause of the intense interest in self-help medical manuals during a time when medicine was, at best, primitive.

Governal's *In This Tretyse That Is Cleped Governayle of Helthe* (1479) is the earliest extant printed English example of this kind of book [31]. As was mentioned in Chapter 3, formatting in these early practical books was often minimal.

Governal's *Tretyse*, perhaps the first printed English medical self-help book, uses rude forms of headings. One, which reads, "¶The Spyces of exercyse," introduces a discussion of the value of exercise:

> Spicies of exercyse ben there well many as there be dyverse states of persons some be strong and some be febel some ryche and some pore some prelates and at the large and some subgettes & enclosed. And somtime wether is fayr and clere/and somtyme not so but derke and reyn/ And therfore it nedeth to hasne species of exercyse/for which the best spyce and the first is to walke to fore mete in high places and clene/ Another specie is to ride and one is for rich men/but great prelates must have other maners of excercise/for when in chamber hall be a great corde knotted in the end and hanged up. and take that corde with both hands and stond upright so that thou touch not the earth and stand a good while/then runne as much as thou mayest heder and thither with that cord [31, p. Aiii].

While the structure of the description moves from the general statement about "species of exercises" to specific examples that become phrased as instruction, the internal organization is weak. As description runs into instruction, once Governal shifts from declarative to imperative, his verbs become more concrete and the v-o relationship more defined. The oral tradition with its emphasis on cumulative phrasing is evident, but the subject-verb relationship within additive clauses linked by conjuncts dominates. The use of *spicies*, a Latin term, indicates that even practical English for lay readers was using Latin borrowings. Otherwise, the diction is marked by predominantly common terms.

The most popular of the medical guides for lay readers continued the plain English, plain style tradition. For example, Thomas Moulton's *This is the Mirour or Glasse of helth* (17 editions, 1531-1580) [32], discussed in Chapters 2 through 4, shows a stylistic maturity beyond the previous 1479 (Governal) example. Clauses are more cleanly separated, and a logical progression is evident in the presentation of ideas. However, there is no rhythmic regularity that would lift the plain style to the middle style. The Elizabethan tendency to use doublets in nouns, adjectives, and adverbs is present, as is the late seventeenth-century habit of capitalizing important nouns, in this instance the names of meats. The following excerpt is from the beginning segment of *The Mirour*:

> The second parte of this Treatyse is howe thou shalte gouerne and preserue thy selfe from the mallice of the Pestilence ayre, and from the venume & corruption that is gendred thereby, the which partye is principall cause of this pestilence that now is reigning. The fyrst and the principall preseruation there agaynst, is this. To gouerne thee wel and wisely, and for to flee all that may gender anye feuer or Age. Fyrst, thou needeth to keepe thee seuerally from all manner of excesse and outrage of meates and drinke, and all manner of feuores meates, as Gose, Doue, Byrds, Malards, Venison, Lampray, Scate and Geles. And also other feuores meats, eate no great meate, no Onyons, nor leekes, nor garlike, nor no fruit [32].

However, when Moulton moves from exposition to instructions for treating specific medical conditions, he shifts to short imperative clauses linked with "and." Diction is characterized by common terms that would be familiar to readers. Doublings disappear in the terseness of imperative mood:

> For a woman Pappe that is sore.

> Take the roote of Bryan, that is to saye, wylde Neppe, and make bare al the ouer part of the sayd rote, and that is to say, the top of the rote, and make an hole therein, and couer it wyth the style stone, and put it in the earth, and lay earth vpon it, and let it stande so foure or fyve dayes, then open it, and take tne juice that ye fynde in the hole and keepe it in a Glasse, and therewith anoynt the brest and it shall be whole [32].

Sir Thomas Elyot's *The Castle of Health*, also discussed in Chapters 2 through 4, was the second most popular self-help medical book and emphasizes that Elyot himself saw that the plain style tradition was best suited to the purpose of instructional discourse [33]. The following description of the gastronomical effects of cucumbers, excerpted from a heavily used 1598 edition of *The Castle*, shows how assiduously Elyot avoided doublets, chose concrete, descriptive terms, and emphasized the subject-verb-object style and clearly demarcated clauses that are predominantly non-periodic. These changes are striking, particularly when we compare *The Castle* with *The Boke Named the Gouernour* [34], discussed at some length in Chapter 3. Mueller argued that Elyot deliberately used a medium-style vernacular in *The Castle*, which unlike *The Gouernour* was devoted to a wide audience. Sententious prose was appropriate for philosophical discourse directed to the English power elite, whereas vernacular medium style was appropriate for explaining and instructing the general population of English readers:

> Cucumbers doe not exceede so much in moysture as Melons, and therefore they bee not so soone corrupted in the stomach. But in some stomaches, being moderately used, they doe digest well: but if they bee abundantly eaten, or much used, they ingender cold and thick humors in the veines, which neuer or seldome is turned into good blood, and sometime bringeth in fevers. Also they abate carnall lust. The seeds as well thereof, as of Melons and Gourds, being dried and made cleane from the huskes are very medicinable against sicknesse proceeding of heat, also the difficulty or let in pissing: they be cold and moyst in the second degree [33].

As Mueller noted, in *The Gouernour*, aureation—characterized by word pairs, neologism, and grandiloquent, solomnic sentences—enhanced the authority of the message [16, p. 255]. A quotation from Section III will emphasize the difference in Elyot's syntax when we compare *The Gouernour* and *The Castle*:

> All be it that some men which have hiderto radde this boke will suppose that those vertues whereof I have treated be sufficient to make a governour vertuous and excellent, nethelas for as moche as the effect of myne enterprise

> in this warke is to expresse, as farre furthe as god shall instructe my poore
> witte, what thinges do belonge to the makings of a perfeyte publike weale,
> whiche well nigh may no more be without an excellent governour than the
> universall course of nature may stande or be permanent without one chiefe
> disposer, and mever, which is over all supereminent in powar, understanding,
> and goodnes [34, Section III, xxiii, p. 268].

In addition to the plain style of general self-help guides, some of the most
pristine examples of early technical description occurred in the uroscopy books
which were among the top sellers during the sixteenth century. In these books, the
authors used metaphor to describe the color of urine. Lacking nomenclature that
accurately characterized urological diseases, writers were forced to use color
descriptors that were as visually precise as possible. As in other technical writing
examples we will examine, metaphor is used only for clarity. For example, urine
is described as "white as clay water of a well," "white as whey," "whytish yellow,
almost as bright as the glow of a lantern," "white russet," "not fully so yelowe as
the yellow apple," "red as safron." In *Here beginneth the seinge of Urynes* (1550),
the following description appears:

> Urine of a woman colored as Lynseed, and she have the flyre then it betokenth
> death.
> Urine of a woman red as gold with a watry circle above, betokeneth that she
> is with child. And take hede if thou se thy face in a womans water, and she be
> with out a fever it betokeneth that she is with chyld. And thou see thy face in
> the uryne of a hot fever, it betokenneth death [35].

The most popular book for midwives was Roesslin's *The Birth of Mankinde . . .*
(10 editions 1540-1604) [36]. Written mainly for women, as noted in Chapter 4,
the style does not suggest that the reader was expected to possess an inferior level
of literacy. Like much medical writing for non-surgeons, it selectively used
medical terminology, but the style was still an unadorned, functional style. The
following excerpt is from a 1598 edition. These opening descriptive sentences
accompany an anatomical drawing of a dissected female torso that reveals the
reproductive and urological organs:

> In this fyrst figure is set foorth the tronke or stocke of a womans bodye, layde
> on the grounde, whose *Peritonium* is opened with the Muscless of the
> Abdomen, and turned ouer toward the inside, according to the use of cutting,
> and afterward we have cut away all the bowels of entrayles from
> *Mesenterium*, the straight entrail being yet left in the body, with also the
> whole *Mesenterium* whose pannicles we have here somwhat taken away,
> and set a portion from the other, that the nature of *Mesenterium* might come
> to light, but this present fygure is for this cause principally declared and set
> forth, that it might so shew the position of the Matrix or womde, and the
> bladder, as in this woman it is seene, no part of the sayde Matrix or wombe
> beyng moved [36].

As was discussed in Chapter 4, the main difference between works for surgeons and medical practitioners and those for midwives was the extent to which Latin medical terminology was used, not differences in syntax. Because the best English doctors studied in France or in Italy, and because the best (most accurate) medical books were translations of French, Italian, or German texts, medical translators, many of them English physicians, quickly adopted the habit of using Latin or French medical terms for which no English equivalent was known. But they often, as the excerpt below shows, added some kind of English equivalent. In the excerpt, from a 1634 book describing surgical instruments by Paré, a renowned French physician, the English translator listed only as H. C., uses doublings—an English and a Latin word—perhaps for instructional purposes for readers with a range of language capability. Thus, the style for technical books for physicians did not deviate from basic plain-style, active-voice clause patterns but used more Latinate diction in an effort to provide precise description:

> For the opening of Tumors.
>
> When a Tumor is ripe, that is, when there is in it *Pus consectum*, at which time all symptomes of heate, paine, and the like are mitigated, and the head of it is growne mucronated and soft, then it is time to open it, and not before. Neither ought the Chirurgian afterward to linger, for feare the humour shut up in the flesh, should contract a venemous, or at least a malignant disposition, and by his contagion, infect the neighbouring parts. And this is especially to be feared, where the adjacent are of more exquisite sense, or of greter dignitie: as also where they are more subject to the confluence of excrements, or in a more prone position; for the weigh of humors will naturally decline. This apertion is made sometimes by a phlegme or Lancet, sometimes by a penknife, as wee call itk, sometimes by Cauteries. The phlegme or Lancet, is that Instrument wherewith is used to open a Veine, and may be of use in tender and soft parts, and where the Apostemation is outward [37].

Medical writing, whether for practitioners or lay readers, also used a plain style, even as early as the middle decades of the sixteenth century. For example, *The questyonary of Chyrurgyens*, published in 1541, exemplifies doublings for clarity:

> Of trouth all the medicamentes and oyntments, the moysteth and heteth more then they and all the grekes call chalaslica, is to say laxatyses, also amonge the saye medycamentes all they that declyne somwhat to drynesse, neuer-theles be not yet clere not manifest contraction (they be called syntatica in greke) such medycamentes ben the remedyes of all ecchymosis. But ye must take good hede if the medycasments applycate to the ruptions and are in depnesse of the body have stronge vertue, and if they be fourer and digestyse, and is resolutyse [38].

Unlike aureate style, medical plain style used parenthetical phrases for definition and amplification. But as in all forms of plain style, each word was functional rather than ornamental. In technical books, we do not see occurrences of

aureate style or indications that writers believed that English needed eloquent augmentation.

Renaissance English Herbals

Although herbals could well be considered in a class by themselves, the better ones, such as William Turner's herbal [39], illustrate the relationship of plant description to medicine. While the early herbals, designed for a general audience, exemplified the same style used by the self-help medical books, Turner's herbal and then Gerarde's (discussed in Chapter 2) illustrate the shift of herbals toward what would emerge as a scientific basis for plant study. While botany as a discipline would not emerge in the Renaissance, and early scientists like Turner had no language for plant analysis or description, Turner's writing illustrates careful, specific use of diction and plain style to describe each plant. Like many early scientists, Turner recognized his debt to Greek science and used its nomenclature whenever he believed it necessary.

Turner is recognized as one of the most vigorous of protestant reformers and the first scientific student of zoology and botany in England [40, pp. 48-49]. For that reason, his use of language is worthy of study in its own right. In Turner's use of plain style for plant description we see his use of the vernacular for classifying plants. He opens each description by providing the names of the plant in Greek, Dutch, French, Italian, and English and then differentiates among the classes of the plant. This excerpt from his description of wormwood illustrates major characteristics of his style:

> But as for astriction or bindings which a man can perceyue by taste is ether verye harde to be founde or ellis none at all. Wherefore Pontike Wormwode oughte to be chosen for the inflammationes of the lyuer. But it hath muche lesse floures and leaues then other wormwodes and the smelle of thys is not onely not unpleasant but resembleth a certeyen spicines or pleasant savor all other haue a very foule smell. . . . Some of later writers leaninge vnto the authorite of Galen . . . that Pontike wormwode differeth muche in kinde from it that groweth In oure counlre euen as Santonike and Sea wormwood do differ. But I for my parte do beleue that they differ in no other wise but that Pontike by the reason of the clyme and complexion of the region where it groweth hath lesse floures and laues than oures hath and for the same cause I beleue that it excelles oures also both in bindings and also in savor or smellinge [39].

Turner worked hard at avoiding metaphor that would suggest a meaning that could be misconstrued by the reader or inferred as ornament. Metaphor was not extended and allowed to control the description offered. Again, we must remember that he was attempting to define and classify plants with no technical nomenclature or knowledge of the internal structure of the plant.

Askham's *A litle herball of the properties of herbes* (1561?), an example of the many cheaply printed herbals available in the Renaissance, well illustrates that precise, functional style was operative in the vernacular before Bacon [41]. Askham described wormwood as follows:

> Wormewode.

> Thys is hote and dry in the seconde degre, it is good for Wormes in the wombe if it be staped, and the joyce wronge out and myngled with swete milk, and geue to the pacient to drinke, a seeth thys herbe in wyne, and make a plaister to the wombe. Make pouder of Wormwode, Centorie, Bettayne, of eche lyke much by weyght, and medle all well together, and the pouder will flee wormes in the wombe, both when it is eaten in potage and drunken. Also for the mylte that is swollen of a colde matter, seeth it in wyne and lette the fyche drinke thereof and that shall heale him, and make a playster of the substance of the herbe, and laye it all hote to thy wombe agaynst the greuace. Also stampe wormewode, and temder it with vynegar, and wyth tested sower break, ground therwith, and with the ioyce of Myntes [41].

Periods are separated by conjuncts, but the sentoids are clearly s-v-o arrangements that progress asymmetrically.

Books on Farming and Animal Husbandry

John Fitzherbert, who was discussed in Chapter 4, wrote a number of popular how-to books on a variety of topics—surveying, farming, and animal husbandry. One of his most popular books was the *Booke of Husbandrie*, which enjoyed 11 editions from 1526 to 1598 [42]. The main differences among editions can be found in the increased information supplied in later editions. The prose was decidedly modern, and the word choice functional:

> How to cure the worme in a Sheepes foote.

> Many times it happeneth among Sheepe, that they haue a worme in their foote which maketh them to halt: take that sheepe and look betweene his clawes, and there you shall find a little hole, as much as a great pinnes head, and therin growth five or sixe blacke haires, about an inch long, or somewhat more, take a sharpe poynted knife, and slit the skinne a quarter of an inche long aboue the hole, and as much beneath and put the one hand in the hollow of the foote vnder the hinder clea, and set thy thombe aboue, almost as the slitte, and thrust they finger vnderneath forard, and with your other hand take the blacke haire by the end, or with the knifes poynt take hold thereof, then pull the haire by little and little, and thrust after thy other hand with thy finger and thy thombe, and there will come out a worme like a peece of fleame, as much as a little finger, and when it is out, put a little Tar in the hole, and it will be quickly wel [42, p. 52].

Note that the purpose of this passage was to enable the reader to perform the process. Metaphor, as in Turner's herbal, was purely functional—a "hole, as much as a great pinnes head" and "a worme like a peece of fleame." Words were used to approximate things rather than to express their intellectual subjectivity. Words as things were a means to an ultimate end rather than artistic expression. The style, in its effort to produce the writer's reality in the mind of the reader, seems to anticipate the style that Spratt believed was the goal of the Royal Society:

> a constant Resolution, to reject all the amplifications, digressions, and swellings of style; to return back to the primitive purity, and shortness, when men delivered to many things almost an equal number of words. They have exacted from all their members, a close, nake, natural way of speaking; positive, expressions; clear senses; a native easiness: bringing all things as near the Mathematicall plainness, as they can: and preferring the language of Artizans, Countrymen, and Merchants, before that of Wits, or Scholars [quoted in 11, p. 95].

A reader unfamiliar with the scientific discourse that Spratt had in mind might have thought that he was referring to late seventeenth-century books on agriculture and husbandry.

Fitzherbert's fourth book of the *Booke of Husbandrie*, which explained the duties of various household servants, provided a character description of the steward, baker, butler, cook, and yeoman, as well as a profile of the ideal husbandman and his wife. Fitzherbert described the steward, who executed the master's authority in his absence. Note that Fitzherbert used doublings to emphasize characteristics of the steward rather than to adorn the description:

> Therefore considering that his authority is so great, and hath so great charge and confidence committed vnto him, hee ought to be a man that knoweth and feareth God, of a good conscience, constant, faithfull, wise, politique, circumspect, diligent, painful, laborious, sad, and graue, in conuersation: sober and gentle in speech: discrent and prudent in reformation: bearing like favour to all person, ready to heare, not light of credite, an example of vertuous living, a mirrour of good manners, neyther to familar, nor yet to strange, constant in countenance, words, and deedes: glad to please his Maister and Mistris, and loath to offend: and finally, such a one as must thinke earnestly his Maisters profit his profit, and his Mistris loss his loss, his Maisters honour and worship is honesty [42, pp. 133-134].

This type of character sketch, certainly not in the Overburian tradition, emphasizes the moral or psychological profile of the particular character portrayed. The description emphasizes the desired behavior of the steward, but the sketch omits impressionistic detail that suggests the attitude of the writer toward the steward's position. Objective description of qualities frame the sketch of the good steward. However, we can see a definite similarity between this character drawing and that of the backbiter in *The Ancren Riwle* (see pp. 145-146), even though the purpose

of the backbiter character sketch was moralistic and the description of the steward informational.

Other farming books continued this tradition of pristine prose. Gervase Markham, like Sir Thomas Elyot, seemed to recognize that technical writing was a genre that required a style decidedly different from that used in books written for pleasure reading. Markham's technical books reveal the tendency shared by other technical writers to present procedures in a highly truncated, concrete, pointed plain style. For example, Markham's *Cheape and Good Hvsbandry* [43], one of the most popular farming manuals in the Renaissance, used plain style throughout, as the work was primarily a how-to manual on the care of animals:

> Chap. XXIII.
> Of paine in the Kidneys; paine-pisse, or the Stone.

> All these diseases spring from one ground, which is onely grauell and yhard matter gathered together in the Kidnyes, and so stopping the conduits of Vrine: the signes are onely that the horse will oft straine to pisse but cannot. The cure is, to take a handfull of Mayden-haire, and steepe it all night in a quart of strong Ale, and giue it the horse to drinke euery Morning till he be well, this will breake any stone whatsoever in a horse [43, p. 21].

In studying Markham's functional style we should recall that technical books show that these writers were capable of various styles, aureate and non-aureate, and that they chose style based on content and audience. Form, then, followed function in format, page design, *and* in style. Markham was capable of producing an aureate, meditative style for his dream vision. In *The Second and Last Part of the First Booke of the English Arcadia* [44], he abandoned conversational diction and syntax in favor of a languid, cumulative sentence structure:

> But during the time that the Nymph, with her charming melody, added a superstuous fetter to mine already bound vp senses, The Princesse, with a minde, variously ouer-burthened with hope, with feare, with desire, with amazement, and with all the extremest worst of confused passions; fate, infinitely longing for that some-thing, which the more, infinitely, feared would present her with nothing; till in the end, casting her faire eye-sight, from the cloudy cutaines of her aged disguise, shee might discerne *Siluagio*, and diuers others of his fellow Forresters, accompanied with the discrete *Oppicus*; and many other Shepeheards came marching towards her; and behind them, as a man forlore, and euen vnworthy of society [44, p. 2].

Books on Renaissance Technologies

The *Short-Title Catalogue* records approximately two dozen technical writers who wrote only how-to books. Most of these books were on technologies, such as surveying, agriculture, military science, medicine, and navigation. These writers consistently exhibited a precise, impersonal, utilitarian style, even though many occasionally lapsed into a first-person point of view. Books presented as dialogue

between two individuals were common. In these books, the style was obviously conversational.

For example, Fitzherbert's *Here begynneth a ryght frutefull mater: and hath to name the boke of surueyeng and improumetes* contains 42 sections on topics such as the number of acres in a field, the worth of an acre, how many beasts can graze on one acre, the size of parks, what a surveyor should do, how to mark the boundaries of fields and meadows, how to correct surveys [45]. Fitzherbert, relying on his 40 years of farming experience for the knowledge he included in the first edition (1523), frequently lapsed into the first person:

> Whether the lorde may gyue or selle the resy-
> due of his forren woodes/ and what su-che gyfte or sale is worthy
> by the yere. etc. Cap.
> vii.

> This letter is playne ynough / and as me semeth no doubte/ but that the Lorde maye gyue or selle the resydewe of the sayde woodes or sastes / Excepte that a manne haue commen of Estouers /but what that gyft or sale is worthy. it is to be vnderstade and knowen and as me semeth the donee or the byoure / shalbe in lyke cuse as the lorde shulde haue ben if he had not guyen it nor solde it. Than the lorde hath improued him selfe as much wodes and wastes as he can laufully and when he hath gyuen or solde the resydue of that he cannot not improue him self of . . . [45, p. vii].

While the syntax is rugged but asymmetrical, the sentoidal development is fully modern.

An interesting book, one that can be classified under either technologies or navigation, is *The safegard of Sailers, or great Rutter* (9 editions, 1584-1640) [46]. The book provides instructions for reaching destinations, such as Amsterdam, Flanders, the coasts of Portugal and Spain, and coasts of England and Ireland. Using the language of seamen because the work was written for seamen, the book was apparently designed to be used during voyages. As Figure 6 in Chapter 2 illustrates (p. 28), the writer included sketches of coastlines followed by additional commentary that explained what the seaman would observe as he approached a particular coastline. Another excerpt from the book reads as follows:

> Item, upon the Skelling are five steeples, one standing at the east end is sharpe, and about the middle of the land are three steeples, two flat ones, and the highest of all is sharp, and that which standeth on the west end is a high flat steeple, and is called S. Brandatius church: and the Island is 3. leagues in length.
> Item, a sougheast or northwest moone make hie water in Rauster deepe.
> Item, husowinen and the Pase, are distant 24. leagues [46, p. 16].

The style here is clearly modern with its cumulative development, common words, and asymmetrical sentoidal constructions.

Another example: Leonard Mascall, in *A Booke of fishing with Hooke & Line* (1590), used personal pronouns—"I will teach you"—but concentrated each section on precise description [47]. For example, the second part of this work, "A Booke of Engines and traps," compiled technical descriptions and instructions for various buzzard and vermin traps. Writing for a broad readership, Mascall provided a woodcut drawing of each trap and a verbal description to explain how the device worked. Like other kinds of technical writing, the description is highly visual but utilizes common terms to explain the way in which the trap works. Figure 4 in Chapter 2 (p. 24) shows the page describing a rat trap:

> This engine is called the Ratte trappe, or fall, which is made with a thicke bottome borde, and two thinner bordes on both sides, and there is two stauues set fast thorowe the bottome borde, then the fall must be thicke bordand heauie withall, and at the endes thereof must your staves goe thorow easely to fall and rise, which two staves have holes above, which staves must also goe throwe the long bridge above, and at the holes ye must put in two pins to holde up the sayde bridge. Then must ye set fast another staffe in the middest of the fall, with a latch in the toppe thereof loose set to fall up and downe: which latch must haue a string, which string conneth downe to the bridge beneath, with a small clicket fastened thereunto: and the bridge is fastened beneath on the backside or borde, an inch from the bottom, and so it is done. Ye may make them to take water Rats in setting them in the water, in the sides of your ponds and rivers, and bayted with carion, but then ye maust set rowes of short nayles under the fall planke, and those will stay either ratte or other fish, if they goe through it, and put downe the bridge [47, p. 64].

An interesting facet of Renaissance technical writing style surfaces in works in which the writer changed style, shifting from sententious to functional style, within the same work. Walter Gedde's *A Booke of Svndry Dravghtes* [48], which shows how to develop designs for glass windows, opens with a one-page introduction by the author. The style is typically Elizabethan and meditative:

> As the principall beautie, and countenaunce of Architecture, consistes in outward ornament of lights, so the inward partes are ever opposite to the eies of the beholdler, taking more delights in the beauty therof, being cuningly wrought, then in any other garnishing within the same. To which purpose, is set downe in this, variety of draughts, some, ordinary and plain, others, curious & pleasant, and although, it may seeme to those that are expert in glazeing, that some of these draughts are needlesse, being so plaine in vse, not deseruing in this sort to be published, yet notwithstanding here I doe in friendly courtesie admonish, that it is most needefull, giuing choice to the builder, both for price, and draught of worke, which by no vndeerstanding can the Glazier so sensibly demonstrate his freat, as by showing his exampls of draught, for by such show, the builders shall vndeerstand, what to make choice off, for whose ease & futherance only, I haue published this practise of glazeing, knowing the expert maister is not vnfurnished of teses vsuall draughts, though each workeman haue not all of them [48, p. 1].

Note that Gedde used doublets and extended phrasing to make his purpose statement more rhetorically emphatic. However, the passage shows an absence of metaphor even within the interlocking clauses. When the actual instructions begin, Gedde changes his style: it becomes direct and precise, as shown in Figure 29 of Chapter 3 (p. 83). In these instructions, "Directions how to make your Square," note that in the first paragraph Gedde used the same verbal-descriptive style as in the introduction; but in the second paragraph, "The ordering of the Square," he shifted to plain style similar to that used by Fitzherbert. The diction correlates with the drawing and illustrates what Adolph calls the "verbal descriptive"—imperative verbs that portray action.

The final section of Gedde's book, "The Manner, How to Anneile, or Paint in Glas," also used strictly functional English. The text correlates with a drawing of a glass furnace and uses the same concrete descriptive adjectives, nouns, and verbs embedded within a conversational sentence. Gedde uses "you," but other than this effort to personalize the instructions, every word has a specific, functional purpose. We can infer that Gedde's objective was to control the meaning for the reader to allow duplication of effort.

Books on Household Management

According to Hull's bibliographical study of guidebooks for women, at least eighty-five practical how-to books were published between 1508 and 1640; twenty-two were cookbooks of varying sorts [49, p. 37]. Again, we must recall Hirsch's point that many more of these books likely existed but perished from extensive use, as the condition of many extant copies indicates [50, p. 11]. In contrast to other books, these guidebooks used a plain style consistently. For example, Thomas Dawson's *The good huswifes Iewell* (4 editions, 1587-1596) [51] contains advice on the care of horses and sheep, recipes for medicines for people and farm animals, and recipes for meats, desserts, and condiments:

To frie Chickins

Take your chickins and let them boyle in verye good sweete broath a prittye while, and take the chickens out and quarter them out in peeces, and then put them into a frying pan with sweete butter, and let them stewe in the pan, but you must not let them be browen with fying, and then put out the butter out of the pan, and then take a little sweete broath, and as much Vergies, and yolkes of two Eggtes, and beate them together, and put in a little Nutmegges, synamon and Ginger, and Pepper into the sauce, and then put them all into the pan to the chickens, and stirre them together in the pan, and put them into a dish, and serue them up [51].

The style of books such as Dawson's anticipates Bacon (with its use of conjuncts, s-v, or v-o order). We can see the similarity if we compare the style of

Dawson's chicken recipe with an excerpt from Bacon's *The New Atlantis*, where, as MacDonald notes, Bacon is "simplicity itself":

> We sailed from Peru (where we had continued by the space of one whole year), for China and Japan, by the South Sea; taking vith us victual for twelve months; and had good winds from the east, though soft and weak, for five months space and more. But when the wind came about, and settled in the west for many days, wo we could make little or no way, and were sometimes in purpose to turn back. But then again there arose strong and great winds from the south with a point east, which carried us up (for all that we could do) towards the north; by which time our victuals failed us, though we had made good spare of them. . . . And it came to pass that the next day about evening, we saw within a kenning before us, towards the north, as it were thick clouds, which did put us in some hope of land; knowing how that part of the South Sea was utterly unknown; and might have islands or continents, that hitherto, were not come to light [quoted in 29, p. 38].

While Bacon was addressing *The New Atlantis* to a more elite group of readers than Dawson was addressing in *The good huswifes Iewell*, the use of coordinating conjunctions (and, but); the emphasis on common, predominantly monosyllabic words; the use of an additive style—all show that the style Bacon was advocating existed throughout the sixteenth century. The linear style of early Renaissance cookbooks, characterized by imperative sentences linked with "and" and "then," shifted to a more analytical, cause-effect style by the early decades of the seventeenth century as cooking, as well as other technologies, showed evidence of the advancement of knowledge. The persistence of the plain style in household management books can be seen in comparing Dawson's style with an even earlier work, *Here begynneth the boke of keruynge*, published in 1508:

> Thou shalte be butteler and panter all the fyrst yere / and ye must haue thre pantry knyues / one knyfe to square trenchour loues / an other to be a chyppere / the thyrde shall be sharpe to make smothe trenchours / then chyppe your souerraynes brede both and all other brede let it be a daye olde / household brede thre dayes olde / trenchour brede four dayes olde / than loke your salte be whyte and drye / the planer mde of Juory two inches brode and thre inches lsonge / a loke that your salte seller lydde touche not the salte [52].

Note the lack of logical progression between clauses, the use of common nouns and verbs, and the close subject-verb relationship. Using Ong's assessment, this early English household book illustrates a highly oral style, where written discourse serves as a means of appealing to memory to help the reader recall information already learned by oral instruction [53]. Yet, even in a style that works as an aid to memory, the spoken quality of the language surfaces.

BACON'S THEORY OF
DISCOURSE AND TECHNICAL WRITING

In continuing to argue, contrary to both Croll and Jones, that functional, utilitarian, or plain style was entrenched in a large number of printed technical books that were in active circulation before Francis Bacon, I further suggest that Bacon's principal contribution was a theory of applied discourse—given credibility by his political and intellectual stature—that provided a rationale for a style already used extensively in various types of functional discourse.

In arguing for knowledge that was useful and practical, Bacon provided a utilitarian ethic for this style already entrenched in utilitarian discourse long before Bacon's impact in the early seventeenth century. In short, the influence of technical writing, legal prose, Biblical translation, history, household management books, farming books, herbals, and medical self-help books had become a major part of the texture of English prose from which Bacon drew. As a prevailing characteristic in works about important aspects of everyday life, this style was as important as the style of classical authors and contemporary rhetoricians to Bacon and post-Restoration scientists in choosing a style suited to the new science. If we now examine Bacon's views on discourse, we can see how his theories reflected and summarized rhetorical practices we have seen operating in technical books of the English Renaissance.

Bacon was interested in discourse as a practical art that would move people to action. While Bacon's theories are scattered throughout his writings, a number of statements indicate his belief that rhetoric was an art of composition that should emphasize invention, selection, and arrangement more than phrasing. Bacon believed that selection of ideas requires that writers decide what to include and what to reject. Both selection and style should be based on the audience, avoid confusion, and lead readers "to things themselves and the concordances of things" [quoted in 53, p. 135]. Form and structure, therefore, must be functional to deliver knowledge so that words became univocal representatives of reality. The various uses of knowledge determine the particular relationship of author and reader, and this relationship in turn determines the kind of prose. Audience, knowledge, and purpose ultimately determine communication. Structure as well as style should be functional and respond to audience. In *The Advancement of Learning*, Bacon wrote,

> And for all that concerns ornaments of speech, similitudes, treasure of eloquence, and such like emptiness, let it be utterly dismissed. Also let all those things which are admitted to be themselves set down briefly and concisely, so that they may be nothing less than words. For no man who is collecting and storing up materials for ship-building or the like, thinks of arranging them elegantly, as in a shop, and displaying them so as to please the eye [quoted in 53, p. 150].

Bacon believed that the order and arrangement of all discourse, both scientific (instructional) and philosophical, should be determined by audience, purpose, and occasion. In thus perceiving that the structure of discourse should be dictated by these three factors, he took an unequivocal stand unrepresentative of his age, as the chief theorists preferred that prose structure should conform to the classical order, with its exordium, narration, proof, and peroration [53, p. 212]. In stressing functional arrangement, Bacon advocated that composition and division proceed according to the evidence of things, and as they really show themselves in nature, or at least appear to show themselves. This view places him in the same camp with Ramus and even Wycliffe, both of whom believed that order should be immediately revealed in the text.

In stressing that structure depends on audience, Bacon criticized his contemporaries who overvalued introductions, those "who study more diligently the prefaces and inducements than the conclusions and issues." He minimized the value of introductions because they prevent one from coming speedily to the argument [53, p. 144].

Bacon is probably best known for his views expressed in *The Advancement of Learning* where he rejected ornamental style. Of the three distempers of learning, the "fantastical" is characterized by an emphasis upon style rather than upon matter. He attributed preoccupation with style to four causes:

> . . . the admiration of ancient authors, the hate of the schoolmen, the exact study of languages, and the efficacy of preaching, did bring in an affectionate study of eloquence and copie of speech, which then began to flourish. This grew speedily to an excess; for men began to hunt more after words than matter; and more after the choiceness of the phrase, and the round and clean composition of the sentence, and the sweet falling of the clauses, and the varying and illustration of their works with tropes and figures, than after the weight of matter [quoted in 54, p. 52].

Recognizing the importance of substance over embellishment, in a letter to Essex Bacon wrote, "The true end of knowledge is clearness and strength of judgment, and not ostentation or ability to discourse" [54, p. 53]. Style and matter are related, he believed, because effective selection of content leads to its efficient expression. But the style the writer chooses depends on the audience:

> For the proofs and demonstrations of logic are the same to all men; but the proofs and persuasions of rhetoric ought to differ according to the auditors; . . . the application and variety of speech, in a perfection of idea, ought to extend so far, that if a man should speak of the same thing to several persons, he should nevertheless use different words to each of them [quoted in 54, p. 150].

Concerning the structure of discourse, Bacon advocated two methods, the analytic and the systatic. The analytic method proceeds from particular to general. Systatic arrangement would begin with an explicit statement and then proceed to

explain the application [53, p. 138]. As we have seen, Wycliffe in his sermons and a number of writers of technical books—Gedde, Markham, Fitzherbert, and Mascall, for example—used the systatic method, which also aligns itself with Ramist method. As Wallace stated, the writer moves through successive steps in division, "from his most general proposition to his special details; and simultaneously, as he funnels down to the special, he progresses from what is clearest and most evident to his audience, to what is new, obscure, and least evident" [54, p. 139]. Figure 17 in Chapter 3 (p. 66), which illustrates the general to the specific via Ramist partitions, shows Alexander Read applying this principle in defining tumors in his surgical lectures to medical students.

Bacon was unique in advocating systatic structure, but he still believed that both structure and style should be influenced by the audience and occasion. Thus, the form discourse takes is essentially functional: order, pattern, and arrangement are controlled by the purpose of communication. The style of the discourse should exhibit three qualities: clearness, appropriateness, and agreeableness [54, p. 146]. Style should conform to subject matter; and both, to the intellect of the auditors.

While he supposedly advocated the *genus humile*, Bacon also recognized that it was not appropriate for every topic. He himself did not adhere to one style. As Adolph [11, pp. 73-75] and Wallace [54] observed, the later essays are copious in their use of examples and are discursive rather than compact:

> Although the early Essays are often brief, studiously balanced, and aphoristic, it is doubtful that Bacon thus wrote because he had become a disciple of Attic prose; rather he thought of himself as a scientist in morals, collecting observations on conduct and affairs for the observation and criticism of others; he . . . used such a style, not because he was rebelling against Ciceronian elegance, but because he thought the manner appropriate to his purpose and to his subject matter [54, p. 153].

The conformity of style to content is again apparent in Bacon's own comment upon the style of his treatise, *A Discourse Concerning the Plantation of Ireland*, which he described as "a style of business, rather than curious and elaborate" [quoted in 54, p. 150]. As Wallace also observes,

> The speeches that were written for the revels at Gray's Inn betray some of the artificial, balanced periods of Lyly's Euphues; so also do some of Bacon's letters, especially those to Elizabeth and James. Indeed, both letters and speeches stand in the strongest contrast to the free-running conversational narrative of the New Atlantis. In brief, if Bacon's style can be said to be that of the genus humile, it is only in the sense that this manner of writing represents a simplicity, directness, and clarity of diction, syntax, and disposition that Bacon seems always to aim at. Basically, his style is functional, for it reflects his purpose, his material, and his public [54, p. 153].

In short, in trying to win acceptance for his great scientific method of discovering nature's laws, Bacon employed various modes of presentation, and in them all

he seems never to have forgotten the importance of audience in applying elements of rhetoric. Bacon paid little respect to figurative and ornamentative uses of language and instead advocated sensible and plausible elocution where the transmission of useful information was the goal of the work. In attacking Hall for his style in the *Chronicle*, Bacon objected to the drama of the presentation. Where Hall presented history with a rhetorical sensuousness, Bacon believed that objective presentation was in order:

> For in a great work it is no less necessary that what is admitted should be written succinctly than that what is superfluous should be rejected; though no doubt this kind of chastity and brevity will give less pleasure both to the reader and the writer. But it is always to be remembered that this which we are now about is only a granary and storehouse of matters, not meant to be pleasant to stay or live in, but only to be entered as occasion requires [quoted in 11, pp. 71-72].

However, Bacon's objection was not new but similar to Ascham's attack on Hall for his way of recounting an address that Henry V supposedly made to his soldiers before Agincourt. Hall, in stark contrast to what Bacon believed constituted effective prose, had written:

> Welbeloved frendes and countrymen, I exhort you heartely thynke and conceiue in your selues that thys daye shal be to vs all a day of ioy, a day of good luck and a day of victory: For truely if you well note and wisely considre all thynges, almightly God, vnder whose protection we be come hither, hath appointed a place so mete and apt for our purpose as we our selues could neither haue deuised nor wished, whyche as it is apt and conuenient for our smal nombre and litle army, so is it vnprofitable and vnmete for a great multitude to fight or geue battaile in [quoted in 17, p. cxix].

Hall's excessive use of doublings led to hot censure by Ascham, who, like Bacon, rejected sententious English as the appropriate style for history. The stringent style of Ascham's attack on Hall stands in stark contrast to the ornateness of the previous passage:

> If a wise man would take Halles Cronicle, where moch good matter is quite marde with Indenture Englishe, and first change strange and inkhorne tearmes into proper and commonlie vsed wordes: next, specially to wede out that that is superfluous and idle, not onelie where wordes be vainlie heaped one vpon an other but also where many sentences, of one meaning, be so clowted vp together as though M. Hall had bene, not writing the storie of England, but varying a sentence in Hitching schole; surelie a wise and learned man, by this way of Epitome, in cutting away wordes and sentences, and diminishing nothing at all of the matter, shold leaue to mens vse a storie, halfe as moch as it was in quantitie, but twise as good as it was both for pleasure and also commoditie [quoted in 17, p. cxx].

While he clearly had little sympathy with the dominant Elizabethan style that wished to augment English with abundant phrases, swelling sentences, and circumlocutions, Bacon was not a disciples of any one classical school. A Renaissance man by any standards, his style shows vestiges of Plato, Cicero, and Aristotle. His commitment to learning suggests that he was aware of the various styles used in England, but he believed that a native, concise style was needed to convey knowledge precisely. Both Wallace and Adolph have also argued that Bacon was not a disciple of any classical school. The rhetoric and style he chose for his great program, particularly the style of discourse, was designed to reveal this new knowledge accurately. As Zeitlin concluded,

> That the man who took all knowledge to be his province, who at the age of sixteen conceived the idea of overthrowing Aristotle and renovating human science, who carried himself with an air of assured lordliness toward the greatest intellects of his day, who enjoyed invincible confidence in the originality and superiority of his genius,—that such a man should, in his maturity, during the busiest years of an active life, put himself to school in order to set down the results of personal observation is, on the face of it, unlikely [55, p. 497].

Adolph believed that to Bacon, the best style was no style at all, and his lack of a systematic theory based on classical models may explain why critical work on Bacon's style is minimal [11, p. 76].

RENAISSANCE TECHNICAL WRITING AND THE USES OF PLAIN STYLE

In tracking the sustained tradition of English plain-style vernacular as it appeared in English Renaissance technical writing, I have attempted to show that too much emphasis has been given to both science and Bacon as the progenitors of this style. Douglas Bush came close to correcting this perception when he wrote that

> we have only to think of the vast bulk of plain writing in books of travel, history, biography, politics, economics, science, education, religion, and much popular literature. Plain prose was the natural medium for most kinds of utilitarian writing, and most writing was utilitarian. . . . Dryden and his fellows represented a culmination rather than a beginning [56, p. 192].

If my discussion has been convincing, then I will have shown how technical writing can be defined as a specific genre of utilitarian writing that was valued and used by a wide range of English readers. The amount of this writing produced during the Renaissance suggests that its contribution to the rise of modern prose style must be considered.

If we define technical writing as that kind of writing which adapts technology to the user, then we can see that process description and instructions were naturally

inclined toward unadorned statement and direct presentation. Writers who were inclined to use a more Elizabethan style in some sections of their technical books lapsed into a rigid plain style in delivering instructions. Thus, process description was less concise than instructions; and background description, less concise than process description. Thus, as we have already observed, technical writers chose diction according to the audience of the work (Chapter 4), the function of a particular segment of a work, and the context in which the discourse would be used (Chapter 3). Thus, actual instructions were strictly functional, as the writer apparently assumed that the reader would consult the text while attempting the process or else commit the steps to memory. The writer apparently assumed that the reader would consult background material in a more leisurely manner.

The upshot of these observations is this: Renaissance non-literary discourse, and within this genre what we can call "technical" writing, perhaps more so than any other kind of Renaissance writing, maintained a plain style that anticipated and in some cases paralleled Royal Society plain style but did so noticeably at least a century earlier. Thus, it was not science with its attempt to capture truth precisely in words which led to the rise of plain style but an increasingly literate public that needed books written in spoken English for self-enhancement. As Bennett [18, 19] and others [20, 57] have already suggested, an English population increasingly literate—but only in English—undoubtedly had much to do with writers using a direct, forceful style that utilized the language of everyday life. The recognition by writers that the financial success of their books depended heavily on readers being able to use the information was likely the rationale for the embedding of plain style in instructional writing.

In addition, as Wallace concluded, little hard evidence suggests that Bacon definitely influenced later seventeenth century writers [54, pp. 205-227]. Because of the danger of determining influence, we can at best say that technical writing as it existed throughout the sixteenth century illustrated the theory of clear discourse as articulated by Bacon. As is often the case, practice precedes theory. Perhaps Bacon himself had access to many of these how-to books and saw the value of their style as a vehicle for instructional discourse. Technical writing did not seek to capture thought processes and the vagaries of human experience, but the desired product. The reader's understanding was a part of this desired product, the transmission of actions necessary to achieve a desired end.

We have only to examine examples of purely Elizabethan writing, best preserved in the writing of standard literary genres, to see the defining differences in technical writing and literary or philosophical writing. As Adolph noted about Hall, "What we find in Hall is an awareness of the sensuous texture of both life and language as well as the moralizing strain. In Bacon [and for technical writing] these qualities are sacrificed to 'business' " [11, p. 71].

It would be incorrect to imply that the prevalence of plain style in technical writing published before Bacon's *Advancement of Learning* did not have a

classical theoretical basis, as Socrates enlisted plain style (just as Bacon did) for the purpose of teaching. In contrast, the high style was to move; the middle style, to delight. Because the purpose of writing was to teach, Socrates advocated a conversational style that was loose in structure rather than rhythmical or patterned. As Morris Croll summarized Socrates' goal for plain style: "Its idios is that of conversation or is adapted from it, in order that it may flow into and fill up all the nooks and crannies of reality and reproduce its exact image to attentive observation" [1, p. xxx]. In short, we can conclude that Socrates also believed in a monadist view of reality.

The importance of re-creating reality for readers meant to Cicero that the writer would be careful in word choice, careful in use of metaphor, and careful in avoiding embellishments of thought. Metaphor that is suitable should embody the common language of listeners. Diction used throughout discourse should be current and familiar. Vivid description should be achieved by clarity and precision, while persuasiveness depends on lucidity and naturalness. Accordingly, exuberant and inflated language must not be sought after in a style meant to carry conviction. The composition must be steady-going and void of formal rhythm [22, p. 8] while avoiding bombast and superfluity. We have only to examine Hooker's prose to see the contrasting, aureate style. A ceremonial discourse used to convey solemnity, grandeur, and order can be seen in his Ciceronian subordination of lesser points to main points to create sentences that float in great length to produce the force of law. While Hooker used a low diction, his ability to transfer the power of the Latin cursus produced a powerful rhetorical statement. As A. D. Hall noted, "he has an ear for a kind of ceremonial language that permitted him to combine a Malory-like simplicity of diction with occasional Latinate grandeur; but mostly because the superficial artifice of syntax and diction is drawing on a discourse of praise that had existed in many genres for centuries" [10, p. 235].

But it was the moderate, predominantly oral style, not Hooker's ceremonial style, that was seen to enhance the marketability of works in English. As Mueller stated, authors attempted to put down their thoughts in a clear and unornamented fashion almost as if they were speaking: "What we observe, in their continued avoidance of rhetorical excess, is a growing body of original composition in fifteenth-century religious prose that bases itself in the locution and sentence forms of actual speech" [16, p. 92]. And, modern research in oral patterns shows that the noun-verb-noun sentoid found frequently in plain-style prose before Bacon is characteristic of oral language. In the plain style, clausal units are left intact and unaltered to produce asymmetry. The result is full clauses and verb phrases that carry a heavy predicative load. Predication, in turn, tends to presuppose and depend upon sequencing, or arrangement in a set order, whether that of events or steps in a process. Asymmetry, then, is a natural and understandable hallmark of the conjunctive syntax which figures so largely in the open sentence forms of the earliest modern English prose.

Ultimately, this approach to the evolution of modern prose suggests several questions that need further study:

1. Given the substantial numbers of utilitarian works published during the Renaissance, to what extent did scripturalism or the need for a non-aureate style to convey information to a newly literate populace become the driving force behind the evolution of plain style?

2. What subtle changes in plain style can be found in studying herbals and gardening books published during the 1475-1640 period? What changes in style can be found in examining medical works written for physicians and surgeons?

3. What changes in style can be found in estate management books and household management books? How is the growth of knowledge reflected in changes in content and in style?

4. How does the occurrence of doublings differ in technical writing and philosophical writing? How do these differences reflect theories for the existence of doublings as proposed by Jesperson [58], Greenough and Kittredge [59], Baugh [60] and others who have examined the use of doublings?

Answers to these questions and others will further our knowledge of Renaissance technical writing and its relationship to current theories of the rise of modern English prose.

REFERENCES

1. M. Croll, Attic Prose: Lipsius, Montaigne, Bacon, in *Style, Rhetoric, and Rhythm: Essays by Morris Croll*, J. M. Patrick, R. O. Evans, M. J. Wallace, and R. J. Schoeck (eds.), Princeton University Press, Princeton, New Jersey, pp. 167-206, 1966.

2. R. F. Jones, *The Seventeenth Century: Essays by Richard Foster Jones and Others Writing in His Honor*, Stanford University Press, Palo Alto, California, 1951.

3. R. F. Jones, *The Triumph of the English Language*, Stanford University Press, Stanford, California, 1953.

4. J. M. Murray, *The Problem of Style*, Humphrey Milford, London, pp. 5, 55-68, 1922.

5. H. Read, *English Prose Style*, G. Bell & Sons, London, 1928.

6. H. C. Wyld, *A History of Modern Colloquial English* (3rd Edition), Basil Blackwell, Oxford, pp. 148-149, 1936.

7. J. R. Sutherland, *On English Prose*, University of Toronto Press, Toronto, pp. 9-19, 58-78, 1957.

8. F. P. Wilson, *Seventeenth-Century Prose*, Cambridge University Press, Cambridge, England, pp. 5-10, 1960.

9. D. C. Allen, Style and Certitude, *ELH 15*, pp. 167-175, 1948.

10. A. D. Hall, *Ceremony and Civility in English Renaissance Prose*, Pennsylvania State University Press, University Park, Pennsylvania, 1991.

11. R. Adolph, *The Rise of Modern Prose Style*, M.I.T. Press, Cambridge, Massachusetts, 1968.

12. S. L. Fish, *Is There a Text in this Class? The Authority of Interpretive Communities*, Harvard University Press, Cambridge, Massachusetts, 1980.

13. G. Williamson, *The Senecan Amble: A Study in Prose Form from Bacon to Collier*, Faber & Faber, London, 1951.
14. M. Schlauch, *The English Language in Modern Times (Since 1400)*, Panstwowe Wydawnictwo Naukowe, Warsaw, pp. 30, 34-37, 111-121, 1959.
15. I. A. Gordon, *The Movement of English Prose*, Longmans, Green & Co., London, 1966.
16. J. M. Mueller, *The Native Tongue and the Word: Developments in English Prose Style 1380-1580*, University of Chicago Press, Chicago, Illinois, 1984.
17. R. W. Chambers, *On the Continuity of English Prose from Alfred to More and his School*, Early English Text Society, Oxford University Press, London, 1932.
18. H. S. Bennett, *English Books & Readers 1475-1557*, Cambridge University Press, Cambridge, England, 1952.
19. H. S. Bennett, *English Books & Readers 1558-1603*, Cambridge University Press, Cambridge, England, 1965.
20. L. B. Wright, *Middle-Class Culture in Elizabethan England*, Cornell University Press, Ithaca, New York, 1935.
21. S. K. Heninger, Jr., Tudor Literature of the Physical Sciences, *Huntington Library Quarterly, 32*:2, pp. 101-133; *32*:3, pp. 249-270, 1969.
22. W. Trimpi, *Ben Jonson's Poems: A Study of the Plain Style*, Stanford University Press, Stanford, California, 1962.
23. J. C. McLaughlin, *Aspects of the History of English*, Holt, Rinehart & Winston, New York, 1970.
24. J. Earle, *English Prose: Its Elements, History, and Usage*, Smith, Elder & Co., London, pp. 13-25, 1969.
25. W. F. Bolton, *A Living Language: The History and Structure of English*, Random House, New York, 1979.
26. F. Mosse, *A Handbook of Middle English*, J. A. Walker (trans.), Johns Hopkins Press, Baltimore, Maryland, 1968.
27. J. H. Fisher, Chancery and the Emergence of Standard Written English in the Fifteenth Century, *Speculum, 52*, pp. 870-899, 1977.
28. M. R. Richardson, The Dictamen and Its Influence on Fifteenth Century English Prose, *Rhetorica, 2*:3, pp. 207-226, 1983.
29. H. MacDonald, Another Aspect of Seventeenth-Century Prose, *Review of English Studies, 19*, pp. 33-43, 1943.
30. J. Bennett, An Aspect of the Evolution of Seventeenth-Century Prose, *Review of English Studies, 17*, pp. 281-297, 1941.
31. Governal, *In This Tretyse That Is Cleped Governayle of Helthe*, London, 1479 [STC 12138].
32. T. Moulton, *This is the Mirour or Glasse of helth*, London, 1539 [STC 18214].
33. T. Elyot, *The Castle of Health*, London, 1595 [STC 7656].
34. T. Elyot, *The Boke Named the Gouernour*, London, 1531 [STC 7635].
35. *Here beginneth the seinge of Urynes*, J. Waley, London, 1550 [STC 22159].
36. E. Roesslin, *The Birth of Mankinde, Otherwyse Named the Womans Booke*, T. Raynald (trans.), London, 1598 [STC 21160].
37. A. Paré, *An Explanation of the Fashion and Vse of Three and Fifty Instruments of Chirvrgery*, H.C. (trans.), London, 1634 [STC 19190].

38. R. Wyer, *The questyonary of Chyrurgyens*, London, 1541 [STC 12468].

39. W. Turner, *The First and Second Partes of the Herbal of William Turner, Doctor of Physick Lately Ouersene Corrected and enlarged with the Thirde Parte Later Gathered and Newe Set Oute with the Names of the Herbes in Grece Latin English Duche Frenche and in the Apothecaries and Herbaries Latin with the Properties Degrees and Natyurall Places of the Same*, London, 1568 [STC 24367].

40. C. E. Raven, *English Naturalists from Neckam to Ray*, Cambridge University Press, Cambridge, England, 1947.

41. A. Askham, *A litle herball of the properties of herbes*, London, 1561? [STC 857].

42. J. Fitzherbert, *Booke of Husbandrie*, London, 1598 [STC 11004].

43. G. Markham, *Cheape and Good Hvsbandry*, London, 1614 [STC 17336].

44. G. Markham, *The Second and Last Part of the First Booke of the English Arcadia*, London, 1613 [STC 17352].

45. J. Fitzherbert, *Here begynneth a ryght frutefull mater: and hath to name the boke of surueyeng and improumetes*, London, 1523 [STC 11005].

46. R. Norman (trans.), *The safegard of Sailers, or great Rutter*, London, 1584 [STC 21545].

47. L. Mascall, *A Booke of fishing with Hooke & Line*, London, 1590 [STC 17572].

48. W. Gedde, *A Booke of Svndry Dravghtes*, London, 1615 [STC 11695].

49. S. Hull, *Chaste, Silent & Obedient: English Books for Women, 1475-1640*, Huntington Library, San Marino, 1982.

50. R. Hirsch, *Printing, Selling and Reading, 1459-1550*, Harrasowitz, Wiesbaden, Germany, 1975.

51. T. Dawson, *The good huswifes Iewell*, London, 1596 [STC 6392].

52. *Here begynneth the boke of keruynge*, Wynkyn De Worde, London, 1508 [STC 3289].

53. W. J. Ong, *Orality and Literacy: The Technologizing of the Word*, Methuen, London and New York, 1982.

54. K. W. Wallace, *Francis Bacon on Communication & Rhetoric*, University of North Carolina Press, Chapel Hill, North Carolina, 1948.

55. J. Zeitlin, The Development of Bacon's Essays, With Special Reference to the Question of Montaigne's Influence on Them, *Journal of English and Germanic Philology*, 27, pp. 469-519, 1928.

56. D. Bush, *English Literature in the Earlier Seventeenth Century* (2nd Edition), Cambridge University Press, New York, 1962.

57. S. L. Thrupp, *The Merchant Class of Medieval London, 1300-1500*, University of Chicago Press, Chicago, Illinois, 1948.

58. O. Jesperson, *Growth and Structure of the English Language* (9th Edition), Basil Blackwell, Oxford, 1948.

59. J. B. Greenough and G. L. Kittredge, *Words and Their Ways in English Speech* Macmillan, New York, 1901.

60. A. C. Baugh, *A History of the English Language* (2nd Edition), Appleton-Century-Crofts, New York, 1957.

From Orality to Textuality: Technical Description and the Emergence of Visual and Verbal Presentation

In teaching technical writing, we often tell our students that graphics—the technique of visualizing information—begins with format and page design, as these aid the reader in accessing information from the page. Once page design is determined, the writer must be sensitive to ways in which drawings, tables, diagrams, flowcharts, and other kinds of pictorial aids can help readers "see"—understand, use, and perhaps remember—the meaning of written discourse. As many of the examples in Chapter 3 reveal, in technical writing, page design and format techniques preceded the use of graphics in printed books. But after the closing decades of the sixteenth century, English writers and printers increased their use of visual aids—particularly tables and woodcut drawings and then copperplates of intricate drawings—to enhance the meaning of the text as well as its visual effectiveness and accessibility. Unlike ornamentation and illumination techniques used in books of prayers, devotions, and poetry, visual design in technical books was functional rather than ornamental, communicative rather than impressionistic. However, in many technical books, technical description produced pages and entire works that presented instructions in an aesthetically pleasing format.

But the emergence of technical writing was dependent upon forces other than advancements in typography. Something else was happening that moved technical writing from general, often inconsistently structured, often dense exposition,

characteristic of most late fifteenth- and early sixteenth-century printed works, to the effectively designed technical books on navigation, medicine, botany, and military science of the late sixteenth and early seventeenth centuries. This "something else" was the combined effects of 1) Ramist-inspired awareness of the importance of visual design in enhancing the clarity of information, 2) the ability to change and control visual design through improving typography, 3) the development and use of distinctive organizational patterns to control the structure of textual presentation of growing quantities of complex information, and 4) the firm establishment of native English syntax as the vehicle for conveying information. All these forces combined to produce one of the hallmarks of technical writing: technical description or the increasingly integrated verbal and visual presentation of objects and concepts captured and molded into text. This emergence, if we trace it in representative technical works of the period, shows how technical writing, by 1640, had emerged into a distinct genre that anticipates technical writing as we know it today.

But a study of changes in the use of visual aids does more than show us that they were used more frequently during the second half of the Renaissance than in the first half. This emergence illuminates the shift from orality to textuality in another form of written discourse besides narrative, the major focus of most orality studies, such as those by Ong [1], O'Keeffe [2], and Clanchy [3]. While Ong states that the oral tradition in literature persisted until the nineteenth century [1, p. 158], the oral tradition in technical writing had been replaced by the textual tradition by the closing decades of the sixteenth century. It is the marked textualizing of technical discourse, its growing separateness from the oral tradition, and its ability to be accessed and understood by an individual reader that makes it a distinctive form within written Renaissance discourse. Technical description becomes a new avenue for tracking the oral-to-textual shift in communication of knowledge.

To understand how technical writing becomes another measure of the emergence of the textual tradition, we need 1) to examine additional reasons for the emergence of technical writing, 2) to examine texts that show the increasing sophistication of technical description, and 3) to assess how this textuality anticipates traditions used in modern technical writing.

WHY TECHNICAL WRITING EMERGED

As explained in Chapter 1, technical writing emerged for three reasons. First, rising literacy created a demand for books in the vernacular, particularly information books that allowed newly literate readers a means of self-education [4]. Second, population growth and proliferation of knowledge through the printed word to a variety of readers meant that knowledge could no longer be passed on orally or depend solely on the oral context to help give meaning to the printed

word. Third, in disciplines such as medicine, expanding knowledge became too cumbersome to be passed on orally. The major shift resulted from the impact of increasing knowledge, empowered by the capabilities offered by steadily improving print technology, which transformed the means by which knowledge was communicated.

Not only knowledge, but also the means by which knowledge was communicated, were crucial. Print technology with its ability to convey information in a variety of verbal patterns and visual forms definitely nurtured experimentation with textual presentation. Seminal thinkers in expanding fields such as logic, rhetoric, and medicine were thus encouraged to use texts to convey their findings to eager readers throughout Europe. Memory became less important as silent reading replaced oral dissemination. Books could be referred to repeatedly. New forms of mnemonic structures were combined with words to aid memorization of textualized knowledge. With the abundance of printed books, more people wanted to know more things than could be shared orally. Writing and printing helped disseminate knowledge to an ever widening circle of newly literate readers who came to depend on the text rather than orally dominated instruction for usable information. Writers soon learned the importance of visual as well as verbal techniques for using the page to convey meaning effectively [1, pp. 299-318; 5, pp. 127-135]. The profit motive led writers in all fields, not just technical writers, to find ways to present information to a non-elite audience who would not be interested in sustained analytical reading.

During the Renaissance, London became famous for its schools, and urbanization helped to fuel the intellectual awakening of the newly literate readers who viewed education as a means of upward mobility [6, pp. 43-80]. Education, no longer limited to formal institutions such as grammar schools and universities, was available in a wide variety of practical short courses and how-to manuals. As stated in Chapter 2, Edmund Howes's *The Third Vniversity of England* (1615) recounts lectures at Gresham College in such practical subjects as arithmetic, swimming, military science, stenography, painting, geography, and navigation [7]. Sir Thomas Gresham was one of several English businessmen who became impatient with the entrenched scholasticism of the universities and founded a college on the premise that education should be practical as well as classical. Many books that may be called "technical writing" dealt with these practical topics and were likely used in extension courses as well as for individual study and application. As Wright observes,

> Gresham's ideal was to provide a combination of humanistic and utilitarian learning brought to the level of the intelligent citizen's understanding. A precursor of modern university extension courses, Gresham College was the first of the great institutions devoted to popularizing learning for the benefit of the middle classes [6, p. 65].

Technical Description and the Rise of Textuality

A number of changes in communication occurred when texts replaced oral communication as a primary means of transferring knowledge. In Chapter 3 on the development of page design and in Chapter 5 on the emergence of modern English plain style, we have already seen how belief in a univocal presentation of reality in terms of a precisely verbalized representation of content was a central concern in technical writing, but the evolution of technical description requires that we see how shifts in both page design and style were part of the emergence of the textual tradition. Technical description—which merges word, format, and picture—became, in the late Renaissance, the hallmark of the technical writing and the textual tradition.

1. Sound was reduced to space. With the complexity of ideas that needed to be packaged to allow silent reading, space had to be used as effectively as possible. Space ultimately dictated meaning, and meaning became word- and then text-bound [5, pp. 127-135].

2. As Ong stated, writing then restructured thought [5, pp. 126-127]. Space, rather than the natural flow of speech, began to dictate method of presentation. Organization also became controlled by the restrictions of space. Narrative, logical exposition began to be replaced by top-down organization that told readers how to read the text that followed. During this time technical description became an icon by which we can see (literally) how text replaced oral transmission of instruction.

3. Because readers were extricated from the context in which the content arose, increased detail was essential to capture meaning. Detail could be presented visually, verbally, and visually/verbally. The textualizing of knowledge, again, made technical description a type of verbal icon [5, pp. 127, 133].

4. The sense of completeness that technical description ultimately offered gave credibility to the text: the text then *contained* meaning [5, p. 126]. The added dimension of visual description integrated with verbal description sought to capture what text alone could not.

As the following discussion will show, a progression in the use and effectiveness of graphics occurred throughout the Renaissance. Better visual display and page design occurred after 1580 than before 1580, with sustained differences emerging after 1600. Noticeable differences exist in the graphical display in books published prior to 1560 and books published after 1560, when Peter Ramus's visual rhetoric began to influence the visual accessibility of text [1]. Printing and knowledge markedly changed the practice of technical description as it progressed from the Middle Ages to the Renaissance.

Medieval Precursors of English Renaissance
Technical Descriptions

Tracking the emergence of technical description in English Renaissance technical writing requires that we examine examples of early precursors of technical description. Renaissance technical description evolved from visual presentation—drawings used to describe objects—to external descriptions, to verbal/visual description, to visual presentation whose effectiveness was heightened by advances in typography.

Early technical descriptions can be traced to 300 B.C. and to pictures that suggest that anatomy was first taught by reference to the human cadaver. Anatomical graphic art originated during Alexandrian Hellenism and produced semi-schematic illustrations, drawn in bold outline, which supposedly used a naturalistic model to demonstrate exenteration and dissection of the cadaver [8] (see Figure 1). Because copyists made successive errors in copying from the originals and then recopying the copies, the drawings contained numerous errors.

Even twelfth-century through fourteenth-century manuscripts did not improve substantially on these drawings. These medieval drawings were often abstractions, rather than naturalistic drawings, of body parts and functions. Figure 2A shows the early concept of the stomach: a sack suspended by a *ductorium*

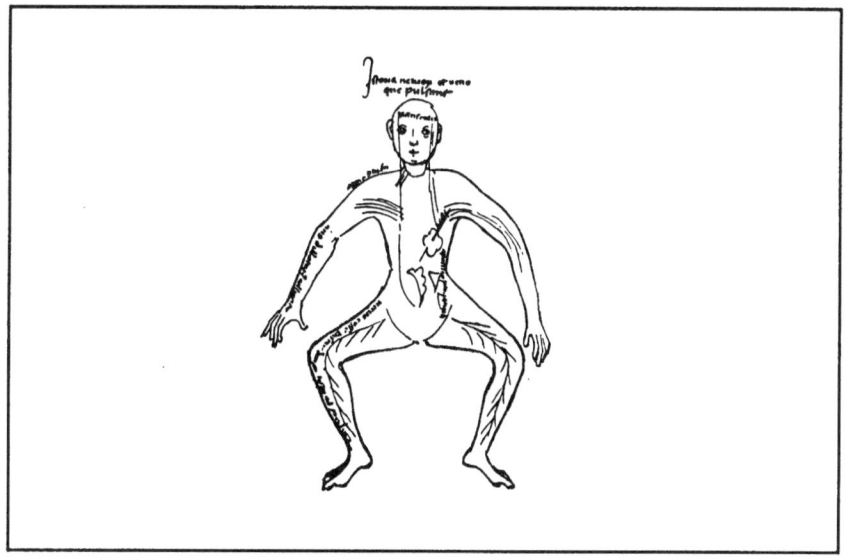

Figure 1. Twelfth-century semi-schematic drawing demonstrating exenteration and dissection of a cadaver [9].

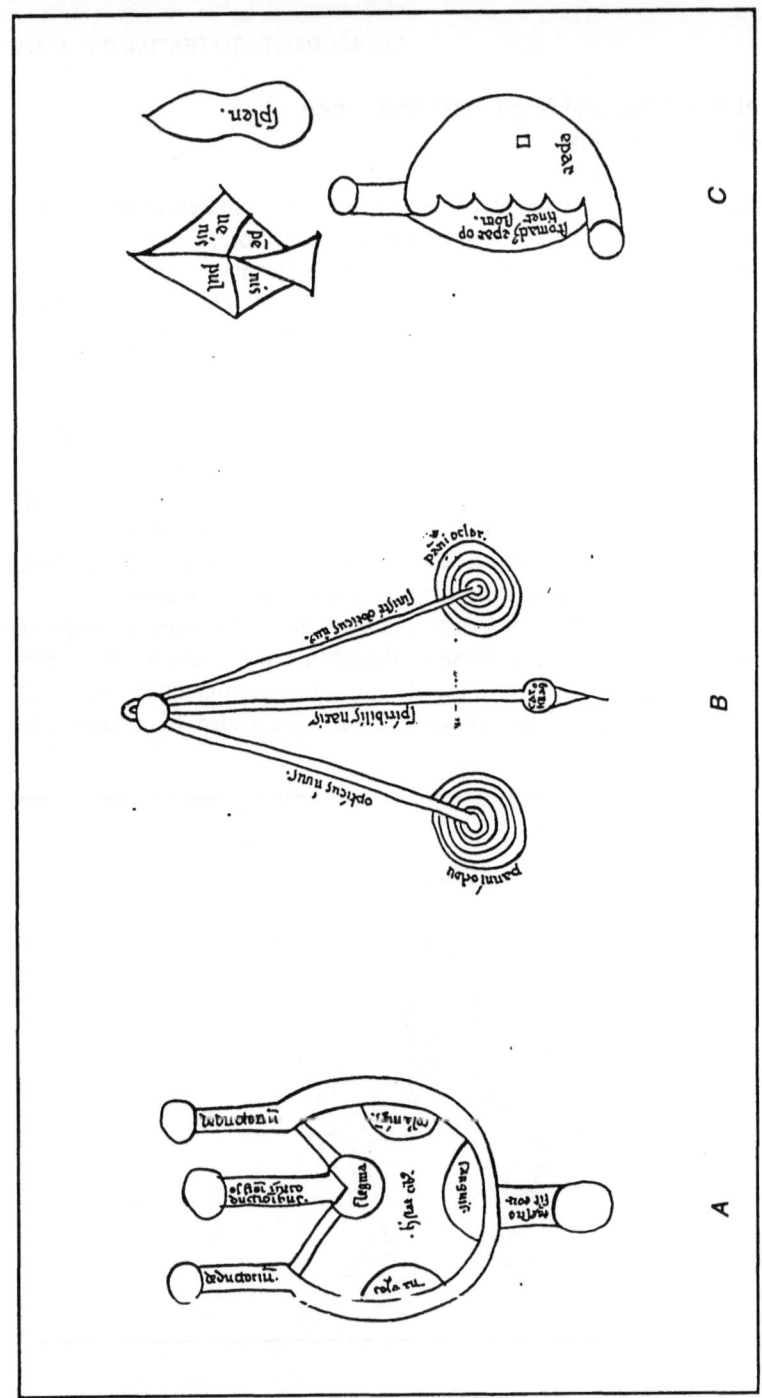

Figure 2. Fifteenth-century incunabula drawings. Representations of A, the stomach; B, the eyes, optic nerves, and nose; C, the liver and stomach, the spleen, and the lungs [9].

surrounding the stomach. The four humors, *flegma, cholera nigra, cholera rubra*, and *sanguinis*, are situated on the stomach wall, at the outlet which leads to the intestine, *egestio fiteorum*. Figure 2B shows a conceptual description of the eyes; their respective nerves, *opticus nervus*; and the nose, *cartillago*, connected to a *spiribilis naris*, perhaps a nerve. All three connect with the brain. Figure 2C represents the liver, with five lobes, as it sits next to the stomach; the spleen; and the lungs, depicted as composed of five angular sections. These perspectives on human anatomy reflected Galenic theory; Galen's theory of anatomy, even though predominantly incorrect, was reverenced until the mid-sixteenth century [8, pp. 40-42; 9, pp. 13-14].

Supporting written discourse, prepared by scribes, was sparse and often used as a means of stimulating the memory to recall concepts presented during oral transmission—lectures and disputation. Figure 3 shows a version of "the wound man" who displays weapons, kinds of injuries, and their causes to serve as a device to prompt the physician's memory. Wound man, a type of descriptive reference graphic, remained an established tradition in medical literature until well into the early seventeenth century. See, for example, Walter Hammond's *The Method of Curing Wounds Made by Gun-shot, Also by Arrowes and Darts, with their Accidents* (1617) [10].

Indication lines that help the reader know on what part of the picture to focus can be traced to the mid-fourteenth century [9, p. 19]. Figure 4 shows two skeletal diagrams, from late fifteenth-century and early sixteenth-century printed books. Both figures show the difficulty that ensued when early illustrators attempted to label drawings by using indication lines extensively. These visuals were typically stiff and lifeless representations, typical of Gothic art. Like early medieval anatomical drawings, they showed a lack of careful examination and precise depiction of the human skeleton.

Lack of realism in description was due to scholasticism, which prohibited dissection of cadavers and preferred abstract, didactic art rather than naturalistic portrayal. In addition, before printing, illustrations had to be simple to facilitate copying. However, illustrations in manuscripts did not serve the same purpose as illustrations in printed books. As indicated in Figures 3 and 4, illustrations were first used to prompt the memory, allowing the medical student to remind himself at a glance of the function of different organs and members of the body. Pictures of structures needed to be clear and not confused and overcrowded with detail.

As illustrated in Figures 1 through 4, what we see most often in medieval manuscripts are outline drawings, accompanied usually by captions which, for example, remind the reader of the places to let blood, the location of different diseases in the body, or the functions of various organs and body parts. Instruction in medical practice occurred by lecture and oral disputation, not by individual reading. Since diseases in the body or the functions of the various organs did not lend themselves to the art of the illuminators, early medical illustrations were

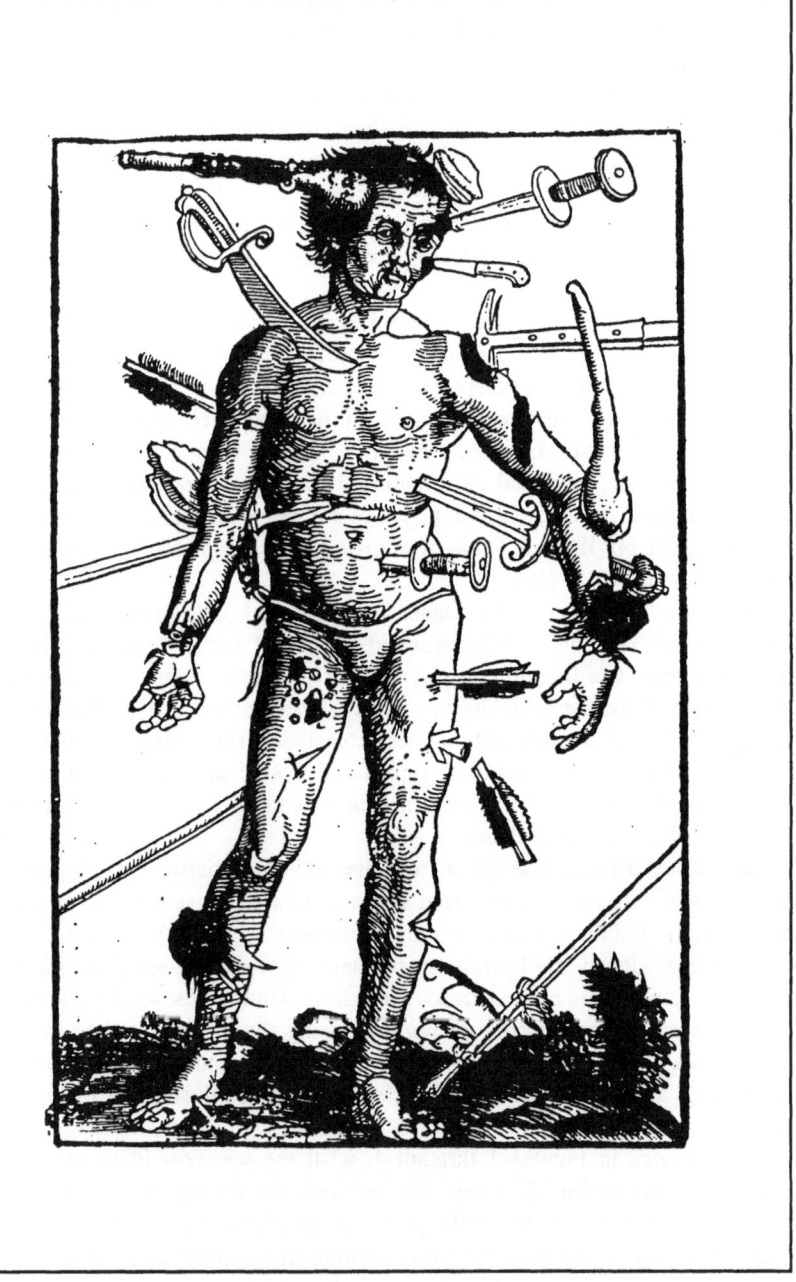

Figure 3. Sixteenth-century medical text drawing,
The wound man [10].

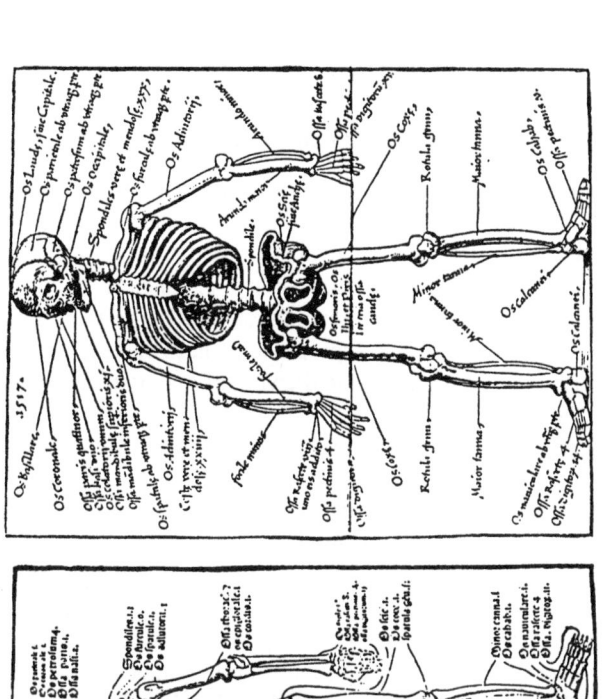
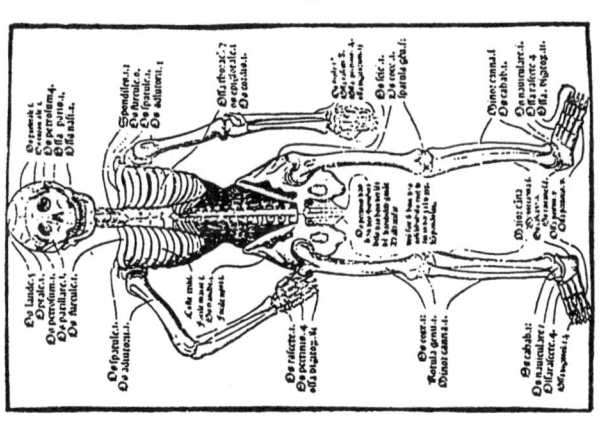

Figure 4. Late fifteenth- and early sixteenth-century drawings. Skeletal diagrams showing the development of the indication line [9].

183

limited in what they could portray. These graphics, we must remember, served as mnemonic devices for helping student auditors remember the lecture.

EVOLUTION OF TECHNICAL DESCRIPTION IN ENGLISH TECHNICAL WRITING

Chaucer's Contribution to Technical Description

Chaucer's *Treatise on the Astrolabe*, written in 1391, supposedly for his son Lewis, exemplifies the best and perhaps the first English technical description [11, 12]. Chaucer's techniques illustrate several qualities that would surface repeatedly in technical descriptions of the sixteenth century: 1) concrete diction within sentoids to describe the astrolabe, 2) drawings of various parts of the astrolabe, 3) placement of the drawings at the appropriate point in the text, 4) integration of the verbal description of a part with the drawing of that part, and 5) reference to the drawing in the text.

While Chaucer was indebted to a Latin version of the *Composition et Operatio Astrolabii* of Messahala, an Arabian astronomer of the eighth century [11, p. vii], the organization of his work, even the development of the introduction itself, exemplifies principles of organization used in modern technical description. In the *Introduction*, Chaucer states the reason he is writing the treatist and then describes the structure, which will include five parts. *Part I* will provide the technical description of the astrolabe. *Part II* will provide instructions for using the astrolabe. *Part III* will be composed of tables of longitude and latitude. *Part IV* will present the theory of the movement of celestial bodies. *Part V* will be a theory of astrology. Although Chaucer completed only the first two parts, these show the relationship of technical description to instructions for using the astrolabe. Chaucer believed that Lewis would better understand procedures for using the astrolabe if he first understood how the mechanism worked. Chaucer's use of technical description to introduce instructions was destined to become a recurring technique in other forms of English Renaissance technical writing, such as medicine, military science, and navigation.

Part I, "Here beginneth the Description of the Astrolabe," is composed of numbered statements that describe each part of the astrolabe. A drawing of that part appears immediately after the descriptive statement. The text beneath the figure reads as follows. Note that the style fully echoes the plain style that persisted in the cross-currents of the rise of modern prose style:

> The east side of thy astrolabe is called the right side, and the west side is called the left side. Forget not this, little Lewis. Put the ring of thy astrolabe upon the thumb of thy right hand, and then its right side will be towards thy left side, and its left side will be towards thy right side; take this as a general rule, as well on the back as on the hollow side. Upon the end of the east line, as I first said, is marked a little cross [+] which is always regarded as the beginning of

the first degree in which the sun rises. From this little cross [+] up to the end of this line, under the ring, thou wilt find the border divided into 90 degrees; and every quarter of thy astrolabe is divided in the same proportion. Over these degrees are numbers, and the degrees are divided into fives as shown by long lines between. The space between the long lines containeth a mile-way. And every degree of the border contains 4 minutes, that is to say, minutes of an hour. {And for more explanation, lo here the figure}: [quoted in 11, pp. 8-9].

Following a drawing of the azimuth, Chaucer uses metaphor and analogy to highlight his description. The text states:

[These] almicanteras are compounded by a or q [or are two degrees apart], but some other astrolabes have the almicanteras divided by one degree, others by two, and others by 3 degrees, according to the size of the astrolabe. The aforesaid zenith is imagined to be the point exactly over the crown of thy head, and also the zenith is the exact pole of the horizon in every region. {And for more explanation, lo here the figure}:

[the figure]

From this zenith, as it seemeth, there comes a kind of crooked lines like the claws of a spider, or else like the work of a womann's caul, crossing the almicanteras ar right angles. These lines or divisions [are called azimuths] [quoted in 11, p. 19].

Extensive variations exist among manuscripts of the *Treatise*, and several include the phrase, "lo here thy figure" to introduce the figure. These phrases are included in brackets { } in the above excerpts to show their location in some editions, such as that of Gunter [11].

Part II focuses on ways to use the astrolabe. Chaucer introduces each use with a descriptive statement. A short paragraph of concise instructions, many with additional drawings, follows each statement. What is particularly interesting about this work is that Chaucer establishes the tradition of describing a mechanism before presenting instructions for operating it, a method used in modern instruction and procedure manuals. Chaucer's method of integrating text and visuals and proceeding to describe the astrolabe according to the spatial arrangement of parts also fully anticipates modern practices.

Leonardo da Vinci's Anatomical and Physiological Sketches

The greatest strides in technical description we owe to Leonardo da Vinci, whose lifelike anatomical sketches and technical inventions, presented with verbal/visual technical descriptions, show the power of combining visual and verbal presentation of detailed concepts. Da Vinci's anatomical descriptions are perhaps his greatest contribution to technical description because these drawings

stand in stark contrast to the unrealistic, rigid illustrations of scholasticism, as exemplified in Figures 1 through 4.

While Leonardo did not publish his anatomical illustrations, compiled between 1505 and 1510, scholars believe his technique influenced Andreas Vesalius, to whom English Renaissance technical descriptions—and in fact all medical instruction—owe an enormous debt. Studies of the sketches indicate that da Vinci planned to write a fully illustrated anatomy book, in which text would be subservient to illustration; that is, anatomical concepts would be presented in lifelike drawings of the human anatomy that used minimal verbal description. That da Vinci was primarily an artist having no formal medical training may well account for his perspective [13, p. 15].

Da Vinci, in his initial studies of anatomy supported by cadavers that he managed to steal and then dissect surreptitiously, soon found the errors of Galenic texts and illustrations. His own work sought not only to correct these errors, but also to use anatomy to advance the study of human beings. Naturalism remains the dominant characteristic of da Vinci's illustrations, which used undershading and white highlighting to portray the vital living body. Comparing Figures 1 and 2 with Figure 5 shows how anatomy, through da Vinci's genius, became animated and multi-dimensional in ways unknown to medieval illuminators.

It is to da Vinci that technical description owes seven significant developments: 1) illustrating different systems of organs in logical succession; 2) illustrating the skeletal system—he was the first to show accurately the bones of the hand [14]; 3) illustrating the relationships between the musculature and the skeleton; 4) showing the relationship between internal organs and the body's surface; 5) using cross-sections to show the topographical relationships between the bones, muscles, and nerves; 6) using exploded views of the skull to show the relationship among its parts (see Figure 6); and 7) using three-dimensional drawings to illustrate structure and function of anatomical parts [9, pp. 70-72; 13, pp. 15-18]. In short, da Vinci provided the visual approach that would become the standard for technical descriptions in less than a century. Figure 5 shows a page of da Vinci's drawings of the hand. His notes for the first frontal drawing of the hand are as follows:

> 27 bones, that is to say: 8 in the wrist (rasette), a b c d e f g h; 4 in the palm K L m n; 15 in the five fingers, i p q r S, o v x y z, 4 7 9 8 6.
> And I give 3 bones to the thumb as to the other fingers, because there are 3 movable body segments like the 3 of each of the other fingers of the hand [quoted in 13].

His notes on the second frontal drawing of the hand translate as follows:

> It is necessary to represent as many bones of these hands as may be separated and distinguished from one another, and with the dimensions and shapes of each bone considered fully from four aspects; and you will note the part of the bone united to those surrounding it, and also the part of the bone not united

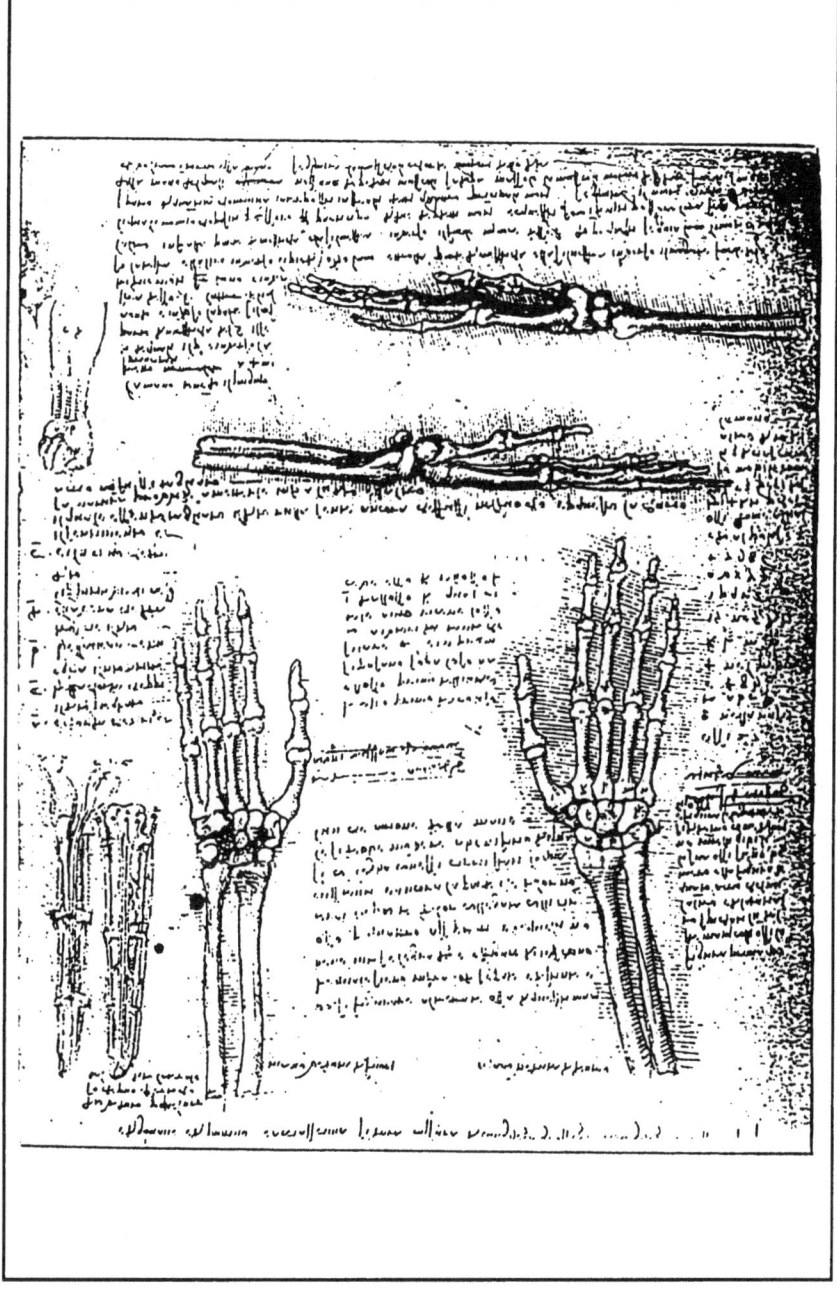

Figure 5. Representation of the human hand,
Leonardo da Vinci (1505-1510) [13].

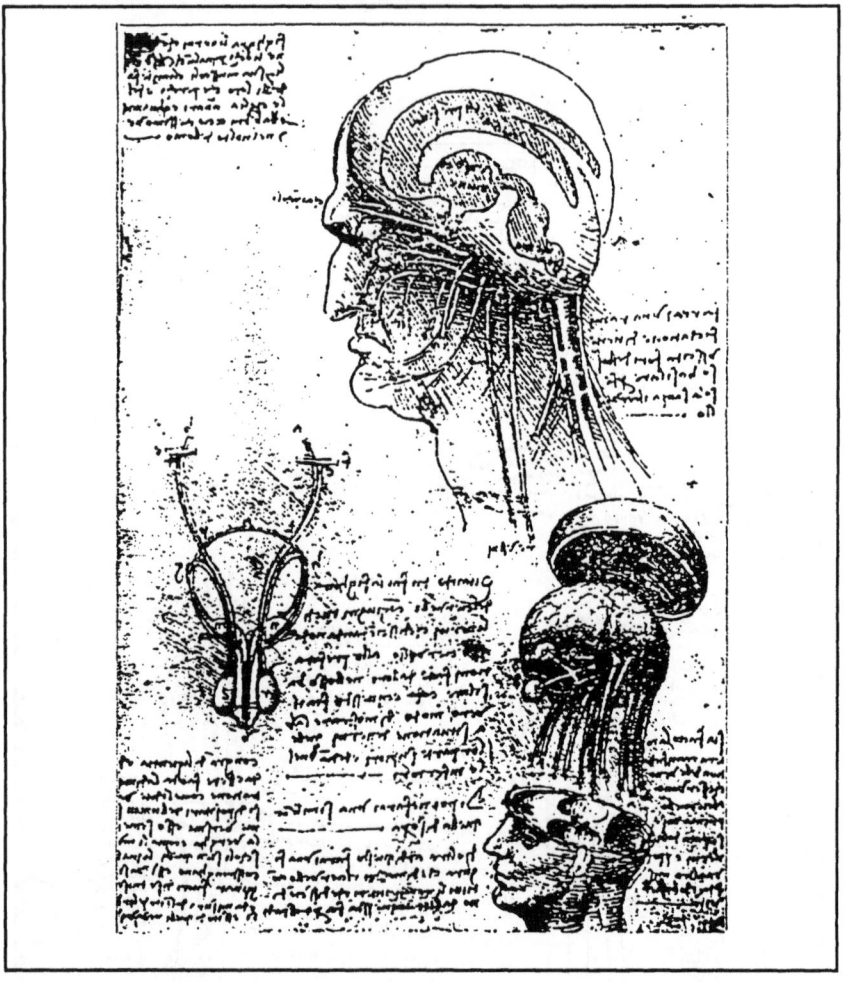

Figure 6. Exploded view of the brain, Leonardo da Vinci
(1505-1510) [14].

with those surrounding it, and which of them must be moved for the service of
whatever action of the hand. The first by one of the thumb and the first bone
of the index finger are placed upon the basilar bone in immediate support, just
as the bone i receives the same support from the bone K that K, receives from
the bone f [quoted in 13, Section 10].

Da Vinci made other contributions to anatomical description. In describing the
muscular system, he suggested a system of nomenclature for the muscles which
would use a separate name for each muscle, a name designed to express its origin,

insertion, direction of pull, and purpose. He also identified the muscles of the face [13, p. 21]. He was the first to describe the heart as a muscle [13, p. 22]. He correctly described the action of swallowing and the course of a bolus of food as it passes over the larynx into the esophagus [13, p. 36]. From an artistic as well as a medieval perspective, he was a seminal contributor to the development of technical description.

Technical Descriptions in Renaissance Anatomy Books

Vesalius's De Humani Corporis Fabrica [15]

Leonardo's most prestigious descendent was Andreas Vesalius, whose *De Humani Corporis Fabrica* (On the Workings of the Human Body) was completed in Venice and Padua in 1542 and published in Basel during the following year [15]. While the *Fabrica* was never published in English, a version of it by Thomas Gemini, *Compendiosa Totius Anatomie Delineatio . . .*, was translated into English by Nicholas Udall in 1553 [16]. This version, as well as a plethora of imitators, brought Vesalius's method of technical description to England and decisively transformed medical illustration and other forms of technical description.

Vesalius combined his commitment to increased verbal detail supported by superb visual illustrations with da Vinci's illustrative methods and fearless rejection of scholastic attitude toward dissection. Both questioned and largely rejected the prevailing reverence for Galen. As a result, Vesalius's *Fabrica* became a milestone in the history of thought and a major precursor of technical description as it exists today. As Charles Singer, a prominent scholar of Vesalius's contributions, notes:

> it is the first great work of science in the modern manner; it is the accepted foundation of the modern disciplines of human and comparative anatomy and physiology; it introduces a sound and positive basis into medical education; some, at least, of its beautiful figures are true works of art and are, moreover, among the triumphs of the woodcutter's craft; it established exact graphic treatment as an essential adjunct of biological research [17, p. xi].

Vesalius's *Fabrica* is vast, a folio of 663 pages illustrated with more than 200 woodcuts. Its imposing layout and spectacular illustrations are immediately impressive, particularly when these are compared with earlier medical works. From a verbal perspective, its language is technical Latin. Its enormous folio pages are dense and devoid of paragraphing. Even though the work is divided into seven books, the density of material hinders easy reference. Despite its awesome but terrifying format and sheer bulk, the work represents the author's contact with the objects he describes. Bodies assume the vitality and movement inaugurated by da Vinci. The result, in contrast to previous medical writings, is a sustained

liveliness. For example, Book VII on the brain (see Figure 7) opens with a series of eighteen figures with extensive, detailed captions that far exceeded the capability of medieval medical writers [17, pp. 90-91]. Da Vinci's and then Vesalius's works clearly demarcate the advancements of knowledge as seen in the concomitant evolution of Renaissance art.

Vesalius's full-figure models also display his commitment to verbal and visual accuracy. Each anatomical drawing is self-contained on its folio page. Parts are labeled, with each part described in the margin. To avoid clutter, Vesalius used letters inscribed on the figure, rather than indication lines. Figure 8, a technical description of the muscular system, is from *The Epitome*, a 25-page work, divided into six brief chapters covering the six main divisions of the human body. Vesalius wrote *The Epitome* in 1543, immediately after completing the *Fabrica* in 1542,

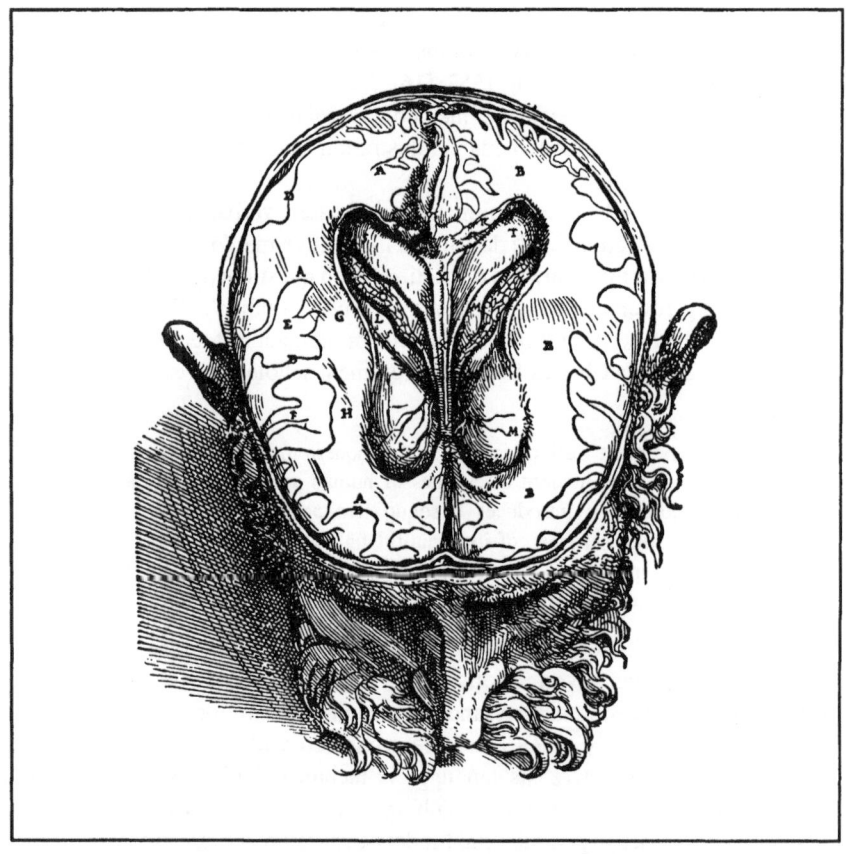

Figure 7. The human brain, from Andreas Vesalius,
De Humani Corporis Fabrica, Libri Septem (1542 edition) [15].

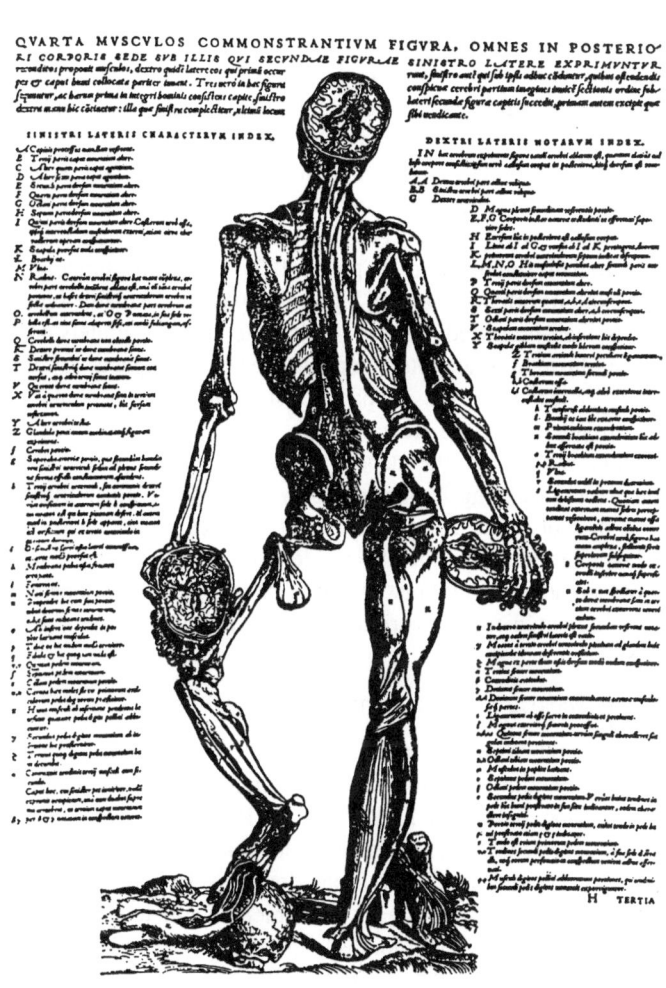

Figure 8. Description of the muscular system, from Andreas Vesalius, *The Epitome* (1543) [18].

and intended the work as both a summary of the *Fabrica* and an introduction to the larger work [18, pp. xxii-xxiv]. *The Epitome* is likely one of the first summaries developed to present the essence of a larger work.

Because detailed technical description was his goal, Vesalius used an organizational technique still prevalent in modern technical descriptions: 1) arranging groups of illustrations of details together at the beginning of each chapter, with extensive legends, and 2) repeating single details throughout the text where repetition had pedagogical value. In addition to full-figure illustrations and illustrations of parts (the brain, urological system, etc.), he used a third kind of illustration: schematic or semi-schematic diagrams built into the text. Each schematic was labeled with a verbal description located adjacent to the illustration [9, p. 110].

In short, Vesalius was the first to use devices introduced by da Vinci and other Italian artists to their full potential [9, pp. 67-102]. Despite its unrelenting density, the *Fabrica* exhibits effective layout, high-quality paper, readable type, an index, and numerous, well-produced illustrations that attempt to accurately depict human anatomy. The illustrations are integrated into the text by explanatory captions and detailed keys or legends—design characteristics that have become standard in technical writing teaching and practice.

To Vesalius is credited the development of books, specifically technical descriptions, for presenting human anatomy accurately, rather than for use in appealing to memory. Because of Italian Renaissance print technology, illustrations could be reproduced without the problem of copyists' errors; and illustration could achieve greater detail [8, pp. 14, 38]. Text and picture became complementary modes of communication, working together to promote understanding for the reader [8, p. 16].

Gemini's Compendiosa Totius Anatomie Delineatio Aere Exarata [16]

English technical writing benefited from those who plagiarized Vesalius, namely Thomas Gemini, who wrote the *Compendiosa Totius Anatomie Delineatio Aere Exarata*, first published in London in 1545 [15]. *Aere exarata* means "engraved in copper"; the illustrations are among the oldest English metal engravings [9, p. 121]. Gemini's effort was sanctioned by Henry VIII, who sought to raise the standards of English surgery, for which he enlisted the support of John Caius, who had lived for several months in the same house as Vesalius [9, p. 122]. Gemini's work was a translation of *The Epitome* of Vesalius.

The *Compendiosa* was a work of 160 pages with forty-three visuals taken from the *Fabrica*. Gemini also inserted numerous blank pages to aid the student in note-taking. The *Compendiosa* became a required text for practicing surgeons, who needed to extend their studies, and for students, many of whom had to study anatomical concepts without having access to a cadaver. Figure 9 shows an excerpt from the technical description of the muscles. Gemini placed the anatomical drawing, its parts labeled with letters, on the page facing the verbal descriptions of the lettered parts. This method of containing a description within facing

pages established a design for technical description that would be used repeatedly in medical literature well into the seventeenth century. The tortured, lifelike stance of the figures exemplifies the application of Baroque art to technical illustration. Clearly, the Gothic tradition in art and technical illustration had passed.

Read's A Description of the Body of Man [19]

Many English doctors, fluent in Latin and educated in Europe, undeniably were influenced by Vesalius and Gemini in designing and writing their own English medical books. One of the most prominent was Alexander Read, whose *A Description of the Body of Man* (1634) uses page layout and anatomical illustrations reminiscent of those of both Gemini and Vesalius [19]. Read's *Description* divides the body into classifications, such as the skeleton, the spine, the bones of the leg, the bones of the hand, the brain, and the circulatory system. Reflecting the technique used by Gemini in the *Compendiosa*, Read's drawings of each group within a classification appear on one page with the technical description on the facing page. As in Vesalius's *Fabrica*, anatomical parts are lettered and illustrations do not use indication lines (see Figure 10, p. 196).

Vesalius's contribution to technical description was timely in the shift to textuality. Printed instruction manuals, such as those by Gemini and Read, allowed medical students the opportunity to learn for themselves, through silent reading. Dissections (of a single cadaver) were conducted in large lecture settings, without the time and frequency available to medical students today. The idea of each student or a team of students having a cadaver for total dissection was unheard of. As Herrlinger noted in his history of medical illustration,

> The substitute—the illustration—is better than no visual demonstration at all; and in practice it was quite impossible to find enough cadavers to meet the demand for anatomical dissection. This is the reason why textbooks were suddenly in such great demand in the second half of the 16th century [9, p. 122].

Roesslin's The Birth of Mankinde, Otherwyse Named the Womans Booke [20]

The influence of Vesalius, through Gemini, was also felt in medical works written for midwives. *The Birth of Mankinde . . .*, a French work by Roesslin translated into English by Thomas Raynald, enjoyed ten editions between 1540 and 1604 [20]. It was the most popular medical guide published for midwives during the English Renaissance. The work includes nine plates of the female anatomy, excerpted from Gemini. Parts are lettered, and each part is described in pages preceding the drawing. Some of the parts shown in Figure 11 (p. 197) are described in the text as follows:

> A.B C.D.   The inner face of parte of the former seate of *Peritonium*.
> EE   A part of *Mesenterium*, knyttyng the thynne intrayles to the backe. . . .

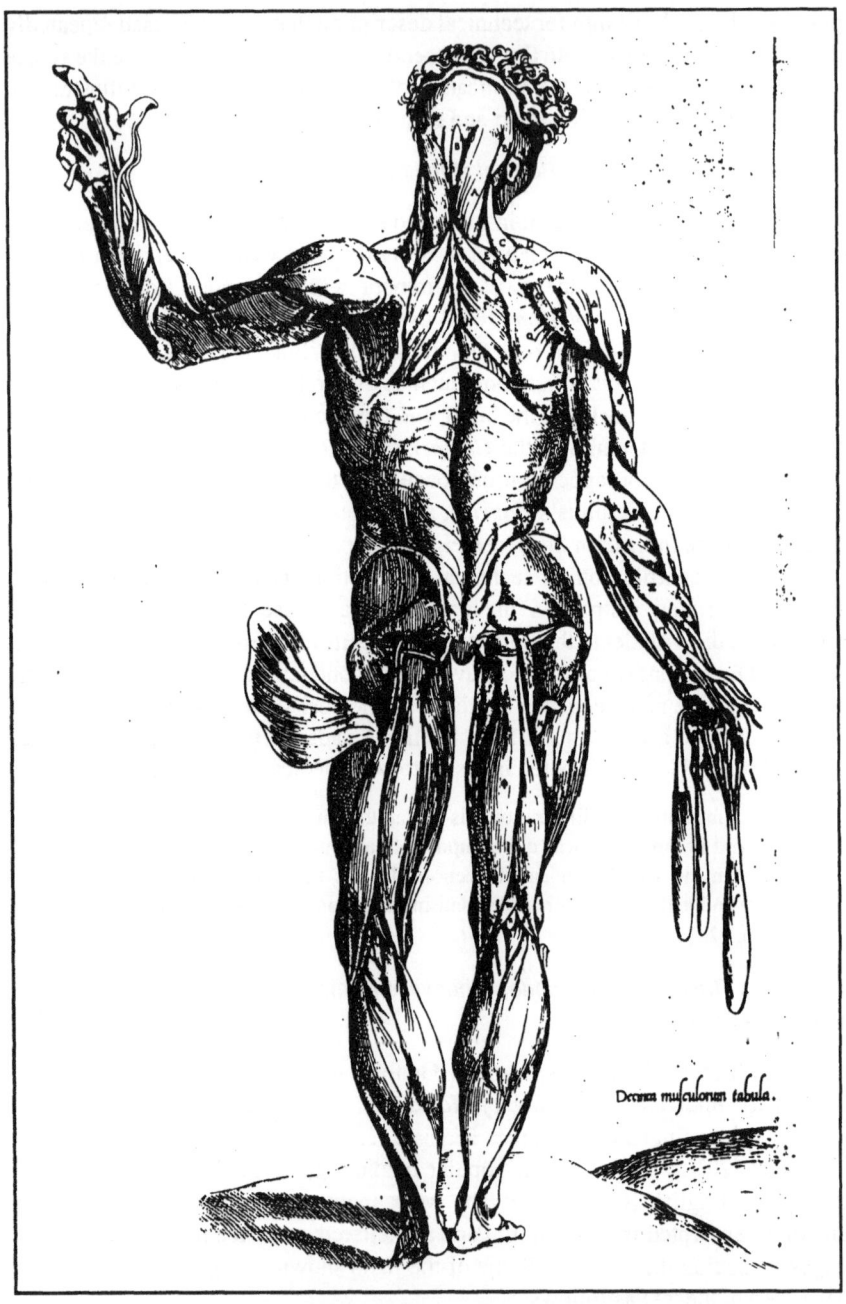

Figure 9. "The Tenth Figure of Muscles," from Thomas Gemini,
Compendiosa Totius Anatomie Delineatio Aere Exarata (1563 edition) [16].

The tenth fygure of Muscles.

from the bone called *Cætix* oz the hyppe bone han=
gynge at the lower feate of the holy bone, where
alfo the lefte mufcle toucheth the ryght, ᵐ ſheweth
the hyghe parte of hys infertion, and ⁿ the lower
parte, whyche becaufe amonge the mufcles of mo=
uynge the ſhynne he ioyneth and deaplye bydeth
hym felfe, euerye parte of hym is not fene.

• The feconde mufcle of moupnge the thigh, bringe
for the mofte parte couered vnder the firft.

ƿ The fyxte mufcle of moupnge the ſhynne, who do=
eth degenerate into that bzoade tendon, whiche is
bewrapped with the mufcles that compaffe aboute
the thighe, but he is not fo thicke but that the muf=
cles vnder hym maye ryghte wel be fene.

ꞅ The ende of the fleaſhye parte, of the mufcle befoze
noted with ƿ.

ʳ The feuenth mufcle of moupnge the ſhynne, coue=
red with the ſhynne couerynge tendon of the fyrte
mufcle of moupnge theſhynne.

ꞅ The fourth mufcle of mouinge the ſhynne, put into
the hynder parte of the feuenth mufcle.

The thyrde mufcle of moupnge the ſhynne marked
in both legges, lyke as certayne other folowynge.

ᵘ A mufcle whych we haue in the place of the fift mo=
upnge of the ſhynne.

ˣ And here appeareth a poztion of the fifte mufcle of
moupnge the thighe.

ʸ The feconde mufcle of moupnge the ſhynne.

ᵗ The firft mufcle of moupnge the ſhynne.

ᵖ The eyght mufcle of moupnge the ſhynne.

ᵧ The nynth mufcle of moupng the ſhynne.

ᴬ In thys bowte the arterye comminge to the ſhanke
with alfo the greate beyne of the legge, and the
greateſt oz thyckeſt ſynowe of all the bodye are con=
ueyed oz caryed.

ᵒ The firft mufcle of mouinge the fote.

ꞎ The feconde mufcle of moupnge the fote.

ᵃ The feuenth mufcle of mouinge the fote.

ᵉ The eyghte mufcle of mouinge the fote.

ᵢ A parte of the leffe bone of the ſhynne and alfo the
outwarde ancle without fleaſhe.

ᵏ Here is fomewhat fene a fmal poztion of the nynth
mufcle of mouinge the fote.

ᵡ The mufcle that bzyngeth the lytle toe from the o=
ther toes.

ᵘ Here is noted a tendon in the left calfe, of mouinge
part of the thyrde mufcle.

ᵥ, The inner ancle.

❧ *The interpretation of the karacters of
the tenth table of mufcles.*

Hys is the tenth in the refpecte of
al the tables befoze, and the feconb
of expreſſyng the backe parte. And
in the order of cuttynge it myghte
wel folow ƿ fourth. In thys do de=
pende certayne mufcles of the ta=
ble befoze, beyng taken away frō their beginnyn=
ges. And here likewife are fene certē mufcles which
in no wyfe ſhewed the felues in ƿ table befoze. And
amongeſt other ƿ mufcle is here taken away, which
in the nynth table we marked wyth Γ Δ bycaufe
he coulde not by hangyng downe in anye place ex= Δ,Ξ,Π,

preſſelye ſhewe him felfe.

ᴬ The righte mufcle of the fyrſte payze of ƿ mouiers
of the heade.

ᴮ·ᴮ The ryght mufcle of the feconde payze of mouinge
the heade.

ᶜ The thirde mufcle of moupnge the ſhoulder.

ᴰ The canell bone.

ᴱ The thirde mufcle of moupnge the bzeſte marked
in the table folowynge with ᶠ.

Γ The fourth mufcle of moiunge the ſhoulder whofe
foure ſydes by roũde befet with *G,H,I,K*.

The fifth mufcle of moupngē the arme.

ᴸ The higheſt parte of ƿ ſhoulder oz ſhoulder poynt.

ᴹ The feconde mufcle of moupnge the arme, whiche

Δ we haue dilygentlye circum:fcribed in the backe
feate of the body yf you do know ꝓ to be infertion oz
knyttyng in of the mufcle foz *T, N & O* do circum=

Ξ,Ο fcribe the endes oz confines of the fayd mufcle lyke
a triangle, hys fourmer parte is fene in the fourth
table of mufcles marked with Ξ.

ꝺ The fyrte mufcle of moupnge the arme.

ꞃ The thirde mufcle of mouing the arme.

ꞏ In thys feat certayn mufcles of moupng the backe
are ſtretched furth, with alfo the fourth of moupnge
the bzeſte.

Θ This mufcle the fecond table ſhewed marked with
Γ, and it is he by whofe benefite ƿ arme is bzought
downewarde towarde the backe beynge alfo the

Ꞅₜ fourth mouer of the fame. *S & T* ſhew the longitude

ᵧ of thys mufcles begynnynge, *V* ſheweth the parte
nexte his infertion, whiche cannot here be ſhewed
but in the feuenth table where he is marked with *O*,

ˣ he is fort. what perceptued, and *X* ſheweth his fide at
that place, where frome the huckle bone he leaueth
to ſpzinge any mooze fourth, the ſydes of this muf=
cle are circumfcribed from *S* to *T* then frome *T* by *X*

ʸ to *V* furthermoze frome *S* to *V*, but *I* ſhall note the
neather angle at the rote of the ſhoulder there ſwel=
lyng forth and couered wyth the thirde ſyde of the
mufcle.

Ζ A poztion of the oblique mufcle of the bealy defcen=
dynge of the Abdome oz Mirach.

ꞏ A mufcle bzyngyng his begynnynge frō the lower
rybbe of the ſhoulder and is the thzuſter furth of
the ſhoulder.

ꞏ A mufcle bzyngynge his begynnyng from the necke
by the heade of the ſhoulder and is the authour of
thzuſtinge furth the cubyte.

ᶜ A poztion of ƿ former mufcle of bowpng ƿ cubyte.

ᵈ A poztion of the mufcle of the bzeſte of bowpng the
cubyte.

ꞏꞏ The longeſt mufcle of puttynge the leſſe bone of
the cubyte vpwarde.

ꞎ The mufcle of ſtretchynge furth the wzeſte with a
forked tendon.

ᵍ In thys feate is fene a thynne couerynge ligament
bynbynge vp the vpper part of the leffe bone of the
cubyte to the ſhoulder.

ᵇ·ᵇ In thys feate the bygger bone of the cubyte is fene
without fleaſhe, oz deliuered from his mufcles.

ꞏ The mufcle of bowynge the bzeſte, whiche is graft
in the ryght bone of the wzeſte.

Thefe.iii.karacters note.iii.beginniges of mufcles
defcendynge frome the bygget bone of the cubyte.

The

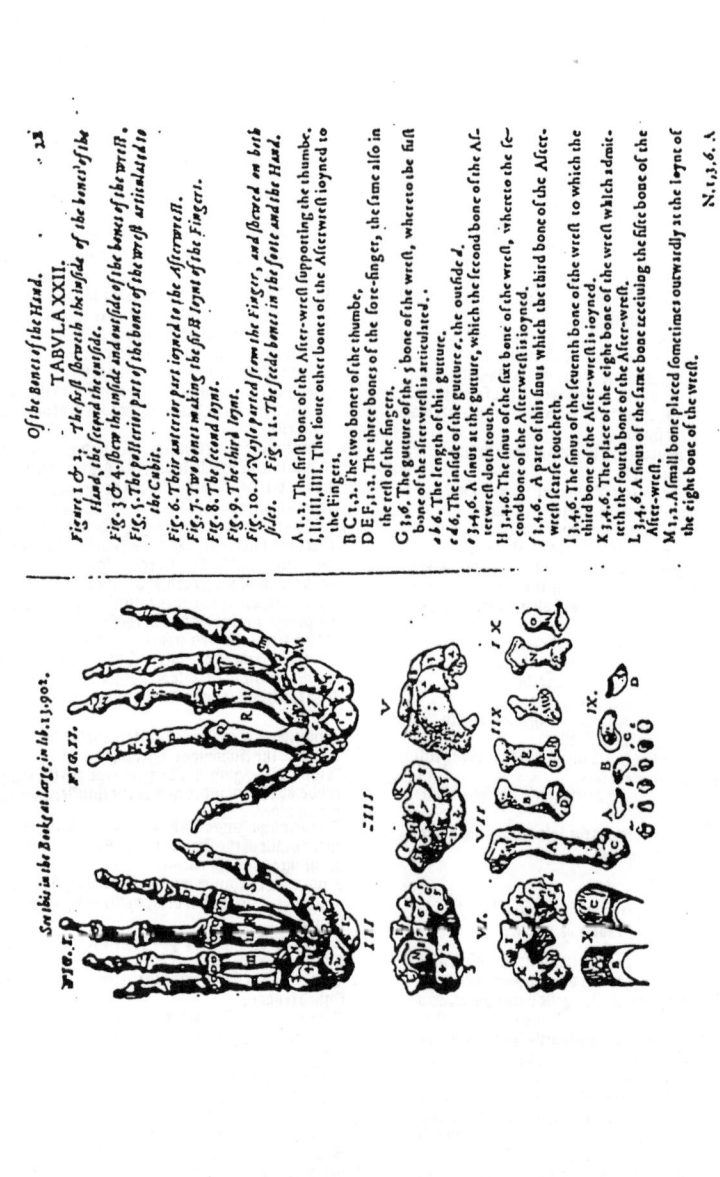

Of the Bones of the Hand.

TABVLA XXII.

Figures 1 & 2. The first sheweth the inside of the bones of the Hand, the second the outside.

Fig. 3 & 4. Shew the inside and outside of the bones of the wrest.

Fig. 5. The posterior part of the bones of the wrest articulated to the Cubit.

Fig. 6. Their anterior part ioyned to the Afterwrest.

Fig. 7. Two bones making the first B Ioynt of the Fingers.

Fig. 8. The second Ioynt.

Fig. 9. The third Ioynt.

Fig. 10. A Naile parted from the Finger, and turned on both sides. Fig. 11. The seede bones in the ioynt and the Hand.

A 1, 2. The first bone of the After-wrest supporting the thumbe.
I, II, III, IIII. The foure other bones of the Afterwrest ioyned to the Fingers.

B C 1, 2. The two bones of the thumbe.

D E F, 1 1 2. The three bones of the fore-finger, the same also in the rest of the fingers.

G 3, 6. The gutture of the 5 bone of the wrest, whereto be first bone of the afterwrest is articulated, 2.

a 6. The length of this gutture.

c d 6. The inside of the gutture c, the outside d.

e 3, 4, 6. A sinus at the gutture, which the second bone of the Afterwrest doth touch.

H 3, 4, 6. The sinus of the sixt bone of the wrest, whereto the second bone of the Afterwrest is ioyned.

f 1, 4, 6. A part of this sinus which the third bone of the After-wrest scarse toucheth.

I 1, 4, 6. The sinus of the seventh bone of the wrest to which the third bone of the After-wrest is ioyned.

K 3, 4, 6. The place of the eight bone of the wrest which admitteth the fourth bone of the After-wrest.

L 3, 4, 6. A sinus of the same bone exceeding the first bone of the After-wrest.

M 1, 2. A small bone placed sometimes outwardly at the ioynt of the first bone of the wrest.

N. 1, 2, 6. A.

Figure 10. "Of the Bones of the Hand," from Alexander Read, A Description of the Body of Man (1626 edition) [19].

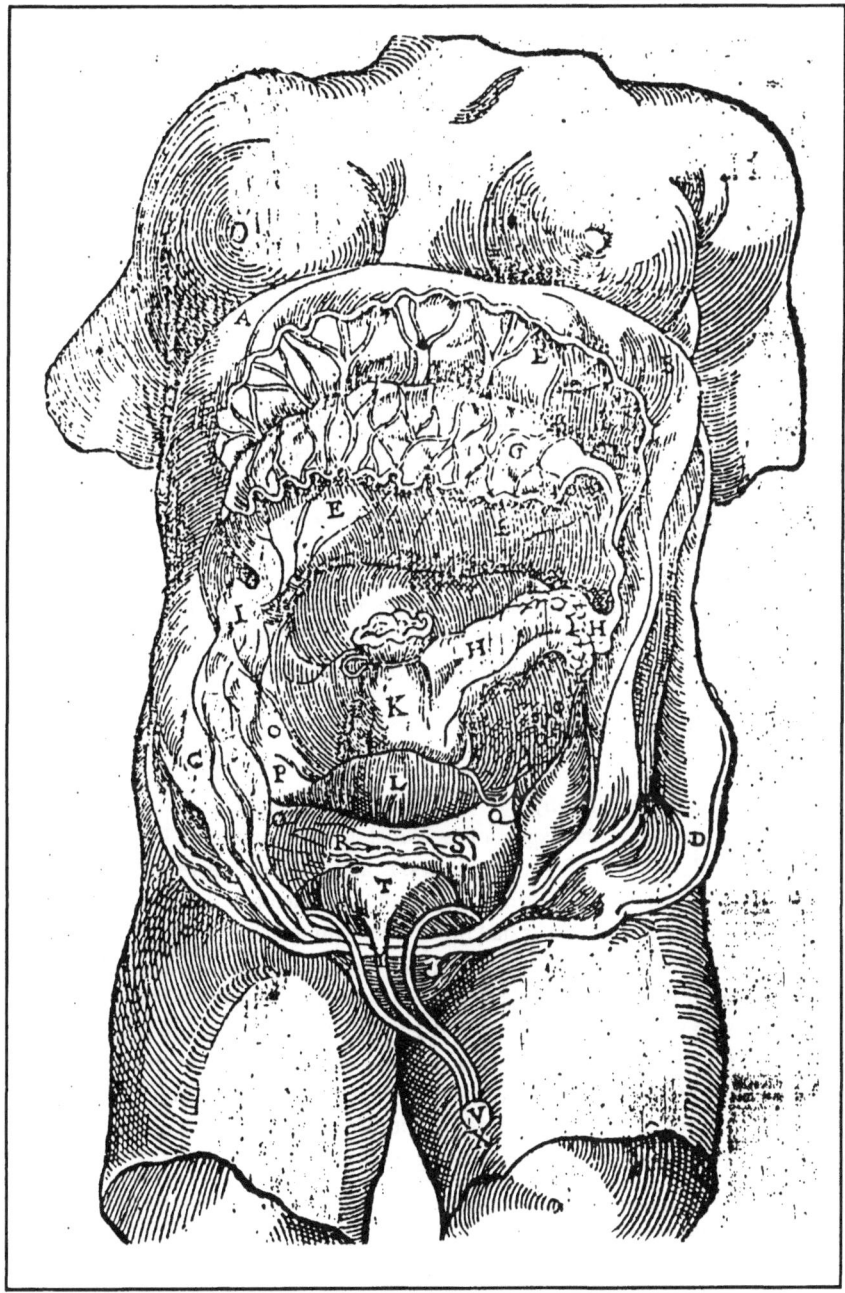

Figure 11. "The First Figure of the Parts of Women," from E. Roesslin,
The Birth of Mankinde . . . (1598) [20].

H H   In this part of *Mesenterium*, the entrayle Colon was committed and set
to, where it was nyett to the straite gut. . . .
K   The strayte gutte beynge there cut of where Colon dyd ende, whiche seate
or place is righte agaynst the place of the knitting together of the holye bone
with the lowest turnynge ioynte of the loynes.
L   The former seate of the bottom of the Matrix, from whence is nothing
perceyuied pluckt away. . . .
T.   The bladder, whose hinder parte is heere chiefly seene. . . .
V.   This is a portion of the Nauell, deliuered in the cutting from *Peritoneum*,
and turned ouer together with the vessels seruing properly to the child [20].

This book, which the preface tells us was written for women, requires a high
degree of literacy, suggesting that women who served as professional midwives
had extensive reading vocabularies, including some Latin medical terminology to
describe the lettered parts of the anatomical drawings.

Technical Descriptions of Surgical Instruments

Technical description also existed in books on surgical instruments, techniques
for handling the equipment, and surgical procedures. The rapid invention of
surgical instruments in the sixteenth century required that these instruments be
described and their use explained through verbal and visual presentation.

Von Braunschweig's The Noble Experyence of the Vertuous Handy Warke of Surgeri [21]

Published in 1525, Hieronymus Von Braunschweig's book on surgical instruments illustrates the visual aspect of technical description before the influence of
Leonardo or Vesalius. As Figure 12 shows, Braunschweig's instruments appear in
shaded, three-dimensional outline. These outline sketches depict tools used in
removing arrows, whose shafts had broken beneath the skin, and various types of
"nypers" or scissors used in cutting flesh. The descriptions lack the detail that
would be used by the end of the century to describe surgical tools. Again, the
descriptions in this work serve as memory prompts, although process descriptions
are given for the use of the trepan. Figure 13 shows the trepan in use. Lack of
realistic presentation in these early sixteenth-century works stands in stark contrast to the technical description used by both Vesalius and da Vinci. The captions
introducing the five tools used trepanning read as follows:

Galen maketh the Trepannes lyke a perser sherp on the end a broder upward
byrauie the instruments in downe pressynge shall not upon dura mater as here
the werth in pycture.

[the figure]

Also the maysters of Parys maketh theyre Trepannes in the maner here
folowynge in figure:

Som stycke fast in the flesshe/ ł som in the bone ł som so depe in the body that it gothe almost through bothe the sydes. And n̄ hā ye be well assured of these/ than it is nede ÿ ye haue som instrument in your helpnes be longynge to this worke. Of the whiche ÿ fyrst be tonges or nyppers/ half mone wyse and inwardly tothed.

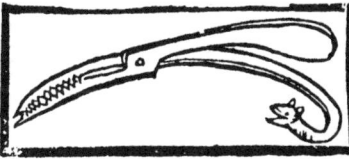

℃ The other be ryht terebellys sharp before/ lyke as ye se fygured and counterfayted here benythe/ whan the hede of the arow is broken of/ thanne ye may ye perce with your instrument in the tymber softely and so drawe it out.

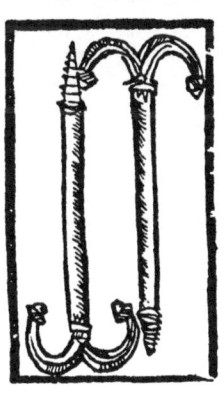

℃ The other be nyppers or prnsors/ namyd Albucasys/ andis formed lyke the byll or nebbe of a storke whin dentyd tethe whiche is fygured as it is shewyde by this instrumēt that stondyth here benethe betwene ÿ other. ij.

℃ The other be tonges or nyppers that be brode/ indentyd and holowe as the nebbe/ for to take out a gone stone/ lyke as here stondyth benethe on the other syde.

℃ The other be terebellys ÿ be bored going in a pipe/ lyke as it stōdith fygured here be nethe on ÿ one syde/ and it is for to take an arow hede out whā the wode is out of it.

Figure 12. Descriptions of surgical instruments, from Hieronymus Von Braunschweig, *The Noble Experyence of the Vertuous Handy Warke of Surgeri* (1525) [21].

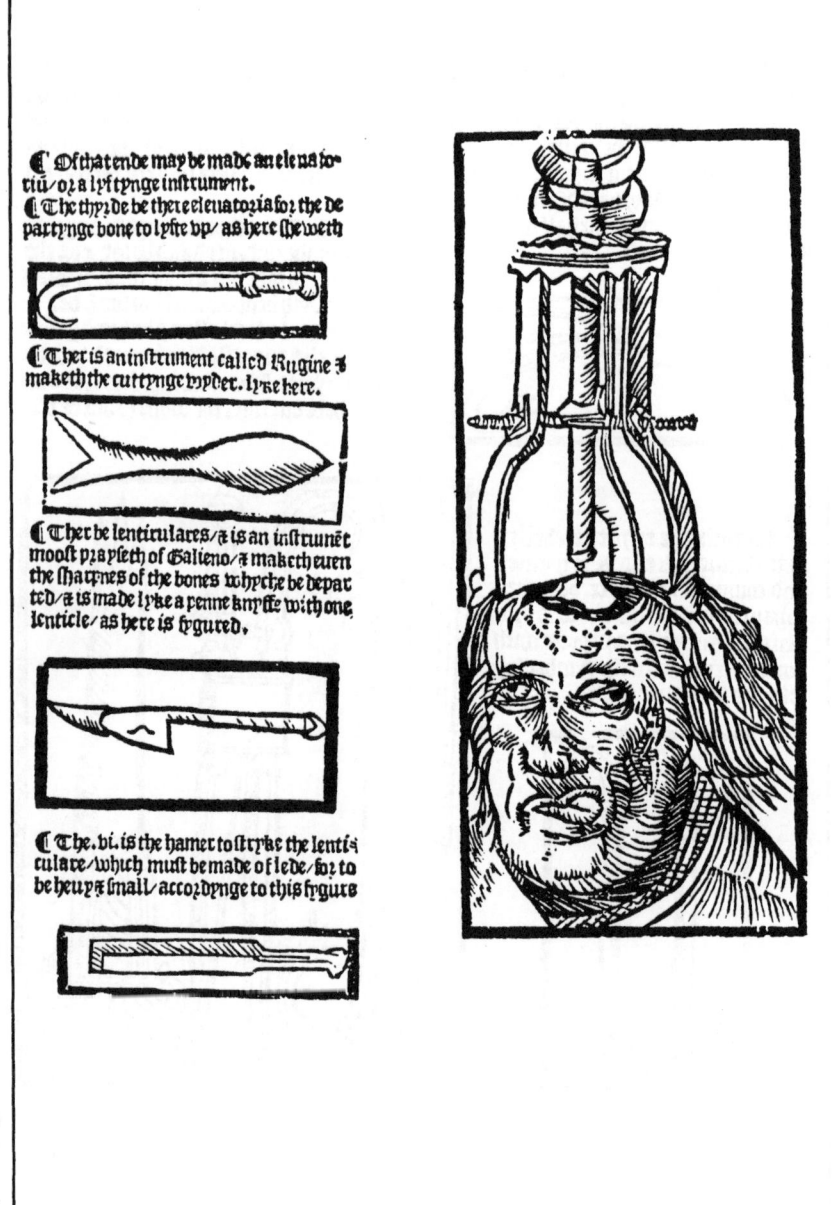

Figure 13. Trepan in operation, from Hieronymus Von Braunschweig,
The Noble Experyence of the Vertuous Handy Warke of Surgeri (1525) [21].

[the figure]

They of Bonony maketh after another maner theyr instrument lyke a sphere.

[the figure]

There is two maners of separatoers or departynge instruments for to separate one hole from the other. the fyrste is Gallicana lyke as here standeth in fygure:

[the figure]

The other maketh of Bonony as here standyth the fygure.

[the figure] [21]

The value of *The Warke of Surgeri* is not its high-quality descriptions of instruments but its descriptions of techniques for using surgical instruments and equipment and its descriptions of surgical operations and procedures. Braunschweig's woodcut illustrating the use of the device to treat a depressed skull fracture (see Figure 13*B*) is considered one of the finest examples of English woodcut of the early sixteenth century [9, p. 142]. Braunschweig's commentary does not describe the device but provides a general process description of its use.

Paré's An Explanation of the Fashion and Vse of Three and Fifty Instruments of Chirvrgery [22]

Ambroise Paré's book, published in 1634, has a much more advanced description of the trepan (see Figure 14) than Von Braunschweig's description in *The Warke of Surgeri* [21] (Figure 13). Paré provided integrated, detailed verbal and visual description. The technology of the trepan had advanced substantially by 1634, as evidenced by the description of each part (Figure 14) given before Paré provided a process description of how to use the instrument.

By the early seventeenth century, technical descriptions of nearly all surgical instruments were more detailed and showed the advances in medicine. Most of the books describing instruments and methods for using them were translations from French medical books. The format of these works—labeled drawings placed directly across from the corresponding verbal descriptions—mirrors Vesalius's *The Epitome* and *Compendiosa*, previously discussed.

Leonardo da Vinci on Plant Morphology

Da Vinci's artistic accomplishments can also be found in his descriptions of plants and gardens, which reflect the same approach he used in his anatomical drawings. He provides a sketch of the herb followed by a description. The format of these drawings and the accompanying notations, developed during 1508-1510, suggest that he was considering a "discourse on herbs," which would have been the first of its kind [23]. The problem he faced was capturing the physical attributes of the plant without scientific nomenclature. He relied solely on

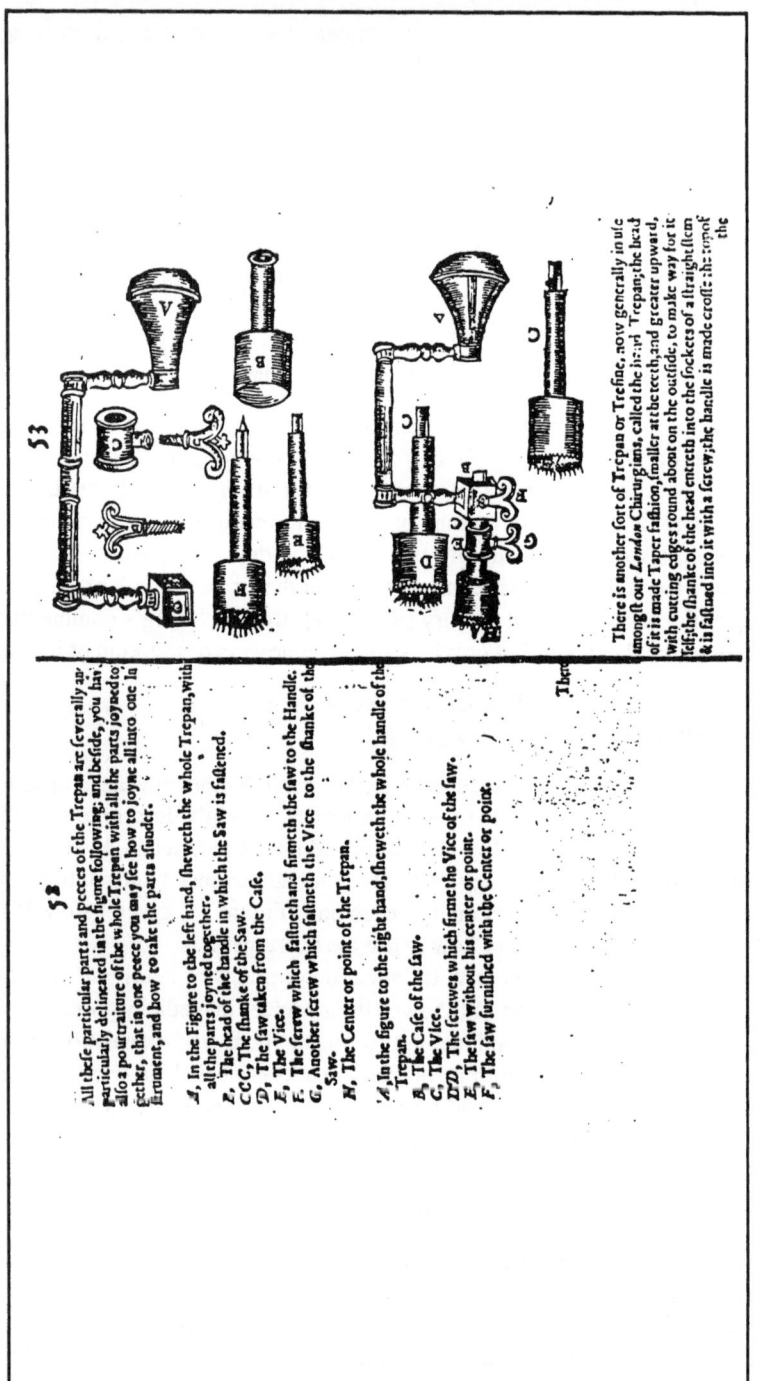

Figure 14. "Another sort of Trepan," from Ambroise Paré, *An Explanation of the Fashion and Vse of Three and Fifty Instruments of Chirvrgery* (1634) [22].

metaphor and analogy to record appearance and size. For example, in describing a rush:

> This is the flower of the fourth species of the rush and is the principal of the kind because it may grow about three to four braccia high, and near the ground it is one finger thick. It is of clean and simple roundness and beautifully green and its flowers are somewhat fawn colored. Such a rush grows in marshes etc., and the small flowers which hang out of its seeds are yellow.

> This is the flower of the third kind that is a species of rush, and its height is about one braccio and its thickness is triangular, with equal angles, and the color of the plant and the flowers is the same as in the rush above [quoted in 23, pp. 148-149].

Botanical Works of the English Renaissance—The Herbals

Botanical works offer another avenue for tracing the development of technical description. Early botanical studies did not suffer from the scholastic prejudice against dissection and close description of plants, as these needed to be recognized in their natural habitat, picked, and then processed for medicinal purposes. It is improvements in illustration technology that account for improvements in illustrations in herbals between 1475 and 1640. However, Vesalius's legacy of detailed description more than likely also influenced herbals. Oddly enough, many respected herbals contained no illustrations. Their reliance solely on verbal description initially enhanced the precision of such description as it was later used in herbals that contained both visual and verbal portrayals of plants.

Von Braunschweig's The Vertuouse Boke of Distyllacyon
of the Waters of All Maner of Herbes [24]

Figure 15, from the 1527 work of Von Braunschweig, shows a rude visual illustration of rosemary [24]. *The Boke* contains no verbal description of any of the plants, emphasizing instead their medicinal values, which are little more than medieval folklore. The woodcuts, exemplified by the cut depicting rosemary, are so rude that the plants appear rigid, lifeless, but typical of Gothic representation. The description of the medicinal values of the water of rosemary (Figure 15) uses a primitive listing arrangement introduced by capital letters:

> A   The same water is good agaynst all colde dysseases/it rectyfyeth and comforteth the spiryt and the natural hete with his good odour/where in the spyryte is rejoysed through his stipysite/wherefore the substance of the membres gather togyther, whan in the mornynge and at nyght is dronke thereof and oure myrced with as moche wyne. B   In the forsayd maner dronke of the same water conforteth and strengteth the braynes/and al other inwarde lymines/the face and the hole bodye wasshed therwith/and the outwarde memnbers rubbed therwith strengtheth the body/and reneweth it and cawseth a man to seme yong and lusty of his corage. . . . D   The same water

Capitulum &c.cv.

Helia ᵹeffari in latyn. The best parte and tyme of theyr dyſtyllacyon is in the begynnynge of June. In the mornynge and at nyght dronke of the ſame water/at eche tyme an oun ce or an ounce and a halfe/is good agaynſte the grauell in the lymnes and in the bladder.

Water of roſemary Ca.cc.xxbi.

Bos marinus in latyn. The beſt parte and tyme of his dyſtyllacyon is/th e leues and the bud dys with the flowres/ſtroped frome the ſtalkes in the tyme of his flowryn. p.iii.

ge dyſtylled/and that may be two ty mes in a yere/but the beſt tyme is in the Mapt. The ſame wa ter is good agaynſt all colde dyſſea ſes/it receyueth and conforteth the ſpiryt and the natural hete with his good odour/where in the ſpyryte is recoyled thrugh his ſtyptycyte/where fore the ſubſtaunce of the members ga der togyder/whan in the mornynge and at nyght is dronke therof an oſi ce myrred with as moche wyne. In the forſayd maner dronke of the ſame water/conforteth and ſtreng teth the braynes/and al other inwar de lymnes/the face and the hole bo dy waſſhed therwith/and the out wardes members rubbed therwith/ ſtrengtheth the body/and reneweth it and cauſeth a man to ſeme yong/& luſty of his corage. In the mor nynge and at nyght dronke of theſa me water/at eche tyme an ounce and a halfe/and the hede therwith enoyn ted therwith/and let drye agayn by hym ſelfe/warmeth the hede/it ſtreg theth and conforteth the wyttes/it cauſeth good memorye and vnderſtā dynge/becauſe it conſumeth ſtegma and melancolye.

The ſame water is good for them that ſwete ſo moche that they become feble and faynt/they ſhall in the mor nynge and at nyght waſſhe the hede and the breſt/than they ſhall be ryd de of the fowle euyll ſwete and ſayn tenes/and come agayne to theyre myght. He the whiche hathe loſt his appetyte/ and hath no luſt nor deſyrynge for to

Figure 15. "Water of Rosemary," from Hieronymus Von Braunschweig, *The Vertuouse Boke of Distyllacyon of the Waters of All Maner of Herbes* (1527) [24].

is good for them that swete so morche that they become feble and faynt/they shall in the mornyng and at nyght wasshe the hede and the brest/than they shall be rydde of the fowle euil swete and sayntences/and come agayn to theyre might [24].

The list also states that rosemary will improve the appetite, warm legs that have poor circulation, reduce swelling in hip bones, open the arteries, prevent pestilence when consumed during a fast, and improve the complexion when applied to the face morning and night.

Turner's Herbal [25]

The rapidity with which technical description developed after the influences of Leonardo and Vesalius can be seen when we examine the work of William Turner, often called the father of British botany. One version of his work, *The First and Second Partes of the Herbal . . . Corrected and enlarged with the Thirde Parte . . .*, represents the first attempt by an English writer to place herbal description on an accurate footing by debunking many myths that surrounded the value of plants [25]. Like other herbalists of his time, Turner knew no relationship among or classification system for plants, but his visual and verbal descriptions are of a much higher quality than Braunschweig's. Despite his effort at precise verbal description, Turner did not seem aware of the need for a technical nomenclature that would enhance the accuracy of his description. He emphasized the difference among varieties of plants, such as the varieties of cucumbers. His descriptions, using common terminology, show the same problems that da Vinci had had in describing plants without a botanical nomenclature:

> The wild cucumbre differeth in nothing from it of the gardine saauinge in the fruyt which it hath no unlyke unto long acornes. It hath leaues and braunches like it of the gardine. The lefe is almost rounde but about the edges full of nickes. The floure is yelow, the fruyte is long, and without there are certeyne long cutters that go from one end to the other and certeine swellinges like rigges where upon growe certeine litle lumpes like ploukes or scabbes. The common kinde of Cucumbre when it is young is grene but when it is ripe it is yelowe. Theophrast writeth in the vi boke and the xiiii chapter that the uttermost parte of a Cucumbre is bitter which thinge as yet may be true so as yet I could neuer finde in those Cucumbers that I haue proued especially when they are ripe. Theophrast and Plinye make three kinds of cucumbres howbeit they do not describe them or tri of any differences in likenes betwene one another [25, p. 176].

Yet in describing other plants, such as rosemary, Turner was unable to proceed much beyond the information included in earlier herbals. His drawings of rosemary (in Part II of his herbal) (see Figure 16) were better than Braunschweig's (Figure 15), but less detailed than those that would be used by Gerarde (see Figure

Of Rosmary.

Jhanotis stephan̄ matike/called in Latine Rosma rinus/is named in English Rosemari. Rosemari (as Dioscorides sayeth) putteth furth smal braunches/ and about them small leues/ thyck/ long/ whyte in vnder/and grene aboue/ with a strong sauor.

The vertues of Rosmary.

Rosmarinum.

Rosinary hath an hetyng nature / Rosinary hea leth ŷ iaundes/ if ŷ broth or water that it is sodde in / be dronken before a man exercyse hym self/ and after that he hath exercised hym self/ entre into a bath/and drynke vnwatered wyne after. Men vse to put Rosmari in me dicines ŷ dryue werisummes away/ and into the oyntment called Gleuci num. The Arabianes as Serapio witnesseth/ gyue these properties vn to Rosinari. Rosinari is hote ŋ drye in the thyrde degre/it is good for the colde reum that falleth from ŷ brayn/ it heateth and maketh fyne or subtil. It dryueth wynde away/ŋ stirreth a man to make water / and bryngeh down weomens floures/ it openeth the stoppynges of ŷ liuer of the milt and the bowelles. Tragus the Ger many writeth that Rosemary is a spi ce in the kitchines of Germany/ and not without a cause. The wyne (sa yeth he) of Rosmari/ taken of a wo man/if she will fast iii. or iiij. houres after/is good for the payn in the mother/ and agaynit the white floures if they come of any inwarde impolteme. It openeth the lung pipes/ and it is good for them that are shortwynded. It helpeth digestion/and withstādeth poyson. It stancheth the gnawyng of the belly/it scoureth the blode/ and if a man will go into a warm bed after that he hath dronken of it/ it will ma ke a man swete. If that Rosemary leues be sodde in wyne/they will do ŷ sa me. The conserue made of the floures of Rosmari / is good for them that swoun /ŋ are week harted. The water of Rosemary as the same Tragus wryteth/ is good for them ŷ for horsenes haue loste theyr speche. Rosemari is also good to withstand trymblyng of the membres/ŋ ŷ disunes of ŷ heade.

Figure 16. "Of Rosemary," from William Turner,
The First and Second Partes of the Herbal . . . (1568) [25].

17). Despite progress in detailed description and classification, Turner's description of the virtues of rosemary remained steeped in folklore:

> Of Rosmary.
>
> Libanotis stephan, matike/called in Latine Rosmarinus/is named in English Rosemari. Rosemari (as Diosondes sayeth) putteth furth smal braunches/and about them small leues/thyke/ long/ whyte in under and grene aboue/with a strong sauor.
>
> The vertues of Rosmary.
>
> Rosmary hath an betyng nature/ Rosemary healeth the jaundes/if the broth or water that it is sodde in be dronken before a man exercyse hym self/and after that he hath exercised hym self entre into a bath'and drynke unwatered wyne after. When use to put Rosmari in medicines they dryue werisumnes away/and into the oyntment called Gleucimum. . . . It dryeth wynde away'and stireth a man to make water and bryngeth down womens floures'it openeth the stoppynges of the liuer of the milt and the bowelles [25, Part II].

However, Turner's text frequently shows his gift for vivid descriptions using similes to help the reader visualize the plant without the aid of technical nomenclature. For example, he compares the dodder (Custuca) to "a great red harep stryng," and the seed vessels of shepherd's purse to "a boyes satchel or litle bagge." Of the dead-nettle he says:

> Lamium hath leves like unto a Nettle, but less indented about, and whyter. The downy thynges that are in it like pryckes, byte not, ye stalk is four-square, the floures are whyte, and have a strong savor, and are very like unto litle coules, or hoodes that stand over bare heades [quoted in 26, p. 154].

It is to these early scientists, like Turner, attempting to work without technical language, that we owe the use of simile and metaphor to enhance the exactitude of instrumental description.

Gerarde's The Herball or Generall Historie of Plants [27]

John Gerarde's massive two-volume, 1500-page folio with excellent woodcut illustrations was not an altogether accurate collection of plant descriptions. However, Gerarde's attempt to provide a complete and useful description of more than one thousand plants makes it a significant work in the developing field of technical plant description. Figure 17 shows Gerarde's woodcut illustration of rosemary [27, Volume I, p. 1008]. In this chapter on rosemary, he included a physical description of each plant, a description of the place where the plant grew, the time of year when it grew, the various names of the plant, the uses of the plant, its negative effects on the body, and its relationship to the humors. In each chapter, each plant is introduced by a heading. The name of the plant appears above the illustration. Each segment of the description is introduced by an italicized heading. Gerarde provided the following description of rosemary:

Of Roſemarie. Chap.6.

✱ The deſcription.

1 ROſemarie is a woodie ſhrub, growing oftentimes to the height of three or fower cubits, eſpecially when it is ſet by a wall: it conſiſteth of ſlender brittle branches, whereon do grow verie many long leaues, narrow, ſomwhat hard, of a quicke ſpicie taſte, whitiſh vnder-neath, and of a full greene colour aboue, or in the vpperſide, with a pleaſant ſweete ſtrong ſmell; among which come foorth little flowers of a whitiſh blew colour: the ſeede is blackiſh: the rootes are tough and woodie.

2 The wilde Roſemarie *Cluſius* hath referred vnto the kindes of Ciſtus Ledon; we haue as a poore kinſman thereof inſerted it in the next place, in kinred or neighbourhood at the leaſt. This wilde Roſemarie is a ſmall woodie ſhrub, growing ſeldome aboue a foote high, hauing hard bran-ches of a reddiſh colour, diuiding themſelues into other ſmaller branches of a whitiſh colour: whereon are placed without order diuers long leaues greene aboue; and hoarie vnderneath, not vn-like to thoſe of the dwarffe Willow, or the common Roſemarie, of a drie and aſtringent taſte, of little ſmell or none at all. The flowers ſtand on the tops of the branches ſet vpon bare or naked footeſtalkes, conſiſting of fiue ſmall leaues of a reddiſh colour, ſomewhat ſhining; after which ap-peere little knaps full of ſmall ſeede. The roote is tough and woodie.

1 *Roſmarinum Coronarium.*
Garden Roſemarie.

2 *Roſmarinum ſylueſtre.*
Wilde Roſemarie.

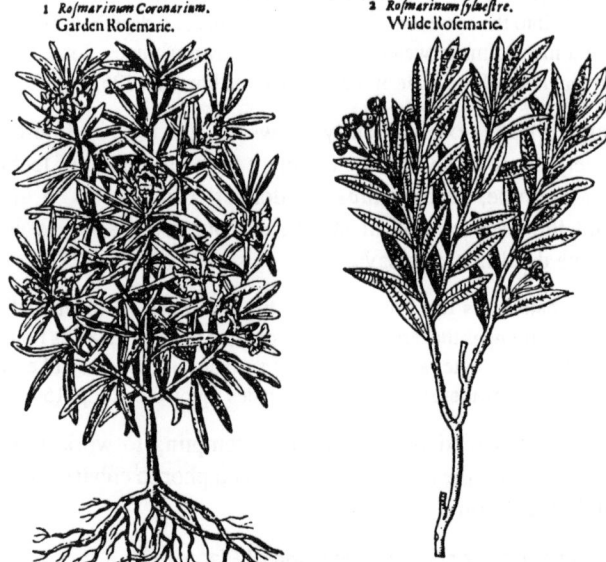

✱ The deſcription.

3 This plant hath ſet ſome controuerſie among the learned, who vndiſcreetly haue confounded *Caſia* with *Caſia*, Canell and Cinnamome. *Anguillara* and *Guillandinus* (two moſt excellent in the knowledge of plants) differ as touching the knowledge hereof, one taketh it to be a kinde of *Lauan-dula*, the other *Roſmarinum Coronarium*. *Theophraſtus* taketh it to be *Caſia*, miſtaking Cammomill for the ſame. Notwithſtanding their controuerſies reſting diſputable, the queſtion may eaſily be decided by the ſimpleſt that haue taken any paines in the knowledge of plants, if theyhad at any time ſeene the plant it ſelfe where it groweth naturally in great abundance, as in *Agro Romano*, and diuers other places, in Narbone, Spaine and Italie. Examine the deſcription who liſt, and they ſhall eaſily perceiue thereby, that it cannot bee *Polygonum Plinij*, neither the Spaniardes *Oſris*. It groweth vp like an hedge ſhrub of a woodie ſubſtance, to the height of two or three cubites; hauing many twiggie branches of a greene colour: whereupon doe growe narrowe leaues like vnto *Linaria*, or Flaxe weede, of a bitter taſte; among which come foorth ſmall moſſie flowers, of a greeniſh yellow colour like thoſe of the Cornell tree, and of the ſmell of Roſemarie: which hath moued me to place it with the Roſemaries as a kind thereof, not finding any other plant

Cccc

ſo

Figure 17. "Of Rosemary," from John Gerarde,
The Herball or Generall Historie of Plants (1597) [27].

fo neere vnto it in kind and neighborhood: after the flowers be paft, there fucceede fruit like thofe of the Mirtle tree, greene at the firft, and of a fhining red colour when they be ripe, like Corall or the berries of *Afparagus*,foft and fweete in tafte, leauing a certaine acrimonie or fharpe tafte in. the end; the ftone within is hard as is the nut, wherein is conteined a fmall white kernell, fweete in tafte: the roote is of a woodie fubftance: it flowreth in the fommer; the fruite is ripe in the end of Oƈtober: the people of Granade, Montpelier, and of the kingdome of Valentia, do vfe it in their preffes and wardrobes, whereupon they call it *Guardalobo.*

Cafia Poetica L'Obelij.
The Poets Rofemarie or Gardrobe.

❋ The place.

Rofemarie groweth in Fraunce, Spaine, and in other hot countries; in woods, and in vntilled places : there is fuch plentie thereof in Langue-docke, that the inhabitants burne fcarce anie other fuell; they make hedges of it in the gardens of Italie and Englande, being a great ornament vnto the fame: it groweth neither in the fields nor gardens of the eafterne colde countries; but is carefully and curioufly kept in pots, fet into the ftoues and fellers,againft the iniury of their colde winters.

Wilde Rofemarie groweth in Lancafhire in diuers places, efpecially in a fielde called Little Reede, amongft the Hurtleberries, neere vnto a fmall village called Maudfley; there founde by a learned Gentleman often remembred in our Hi-ftory(and that woorthily)mafter *Thomas Hesketh.*

❋ The time.

Rofemarie flowreth twife a yeere,in the fpring, and after in Auguft.

The wilde Rofemarie flowreth in Iune and Iulie.   *❋ The names.*

Rofemarie is called in Greek Λιϐανωτις ϛεϕανωματικη: in Latine *Rofmarinus Coronaria* : it is furnamed *Coronaria,*for difference fake betweene it and the other *Libanotides,* which are reckoned for kindes of Rofemarie, and alfo bicaufe women haue been accuftomed to make crownes and garlands there-of: in Italian *Rofmarino coronario:* in Spanifh *Romero:* in French and Dutch *Rofmarin.*

Wilde Rofemarie is called *Rofmarinus fyluefiris,* of *Cordus Chamæpeuce.*

❋ The temperature.

Rofemarie is hot and drie in the fecond degree, and alfo of an aftringent or binding qualitie, as being compounded of diuers parts,and taking more of the mixture of the earthie fubftance.

❋ The vertues.

A   Rofemarie is giuen againft all fluxes of bloud; it is alfo good efpecially the flowers thereof for al infirmities of the head and braine,proceeding of a colde and moift caufe; for they drie the braine, quicken the fences and memory,and ftrengtheneth the finewie parts.

B   *Serapio* witneffeth, that Rofemarie is a remedie againft the ftuffing of the head,that commeth through coldnes of the braine, if a garland thereof be put about the heade, whereof *Abin Mefuai* giueth teftimonie.

C   *Diofcorides* teacheth that it cureth him that hath the yellow iaundies,if it be boyled in water and drunk before exercife, & that after the taking therof the patient muft bathe himfelfe & drink wine.

D   The diftilled water of the flowers of Rofemarie being drunke at morning and euening firft and laft, taketh away the ftench of the mouth and breth, and maketh it very fweete, if there be added therto,to fteepe or infufe for certain daies,a few Cloues,Maces, Cinnamom, & a little Annifeede.

The

Figure 17. (Cont'd.)

Of Rosemarie. Chap. 6

Rosemarie is a woodie shrub, growing oftentimes to the height of three or fower cubits, especially when it is set by a wall: it consisteth of slender brittle branches, whereon do grow verie many long leaues, narrow, somwhat hard, of a quicke spicie taste, whitish vnderneath, and of a full greene colour baboue, or in the vpperside, with a pleasant sweete strong smell; among which come foorth little floers of a whitish blew colour: the seede is blackish: the rootes are tough and woodie [27, Volume I, p. 1008].

Like Turner, Gerarde included a section on "The vertues" of each plant (its medicinal values), but the description of the plant assumes greater importance than it had in previous herbals. Botany's development toward an emerging science can be seen when we compare sample pages from Braunschweig (1527) and Gerarde (1597) (Figures 15 and 17). Note that Gerarde was much more interested in the precise external description of rosemary and the varieties of rosemary, a focus that anticipates the shift toward plant taxonomy. Braunschweig, in contrast, emphasized the folk medicine values of rosemary. Comparing these two works also reveals that graphics were becoming an essential element in supplementing language in the presentation of knowledge. In short, knowledge made advanced typography possible, but typography also became essential for the presentation of expanding knowledge.

The chief purpose of the early herbalist was to enable the reader to identify herbs used in medicine. Thus, until the middle of the sixteenth century, the pictures in the herbals were often so conventional and the descriptions so inadequate that the books must have been relatively useless without the reader's having a basic knowledge of plants, transmitted orally. Because knowledge of plants was still transmitted by oral instruction and demonstration, these early herbals could be used for reference, rather than as a guide to be understood only through private reading.

One difficulty the herbalists faced was that the purpose of the parts of the flower was not yet understood. Thus, the description of the stamens and stigma of the potato as a "pointell, yellow as golde, with a small sharpe greene pricke or point in the middest thereof," as vague as it seems, was the best that the herbalist could do with no knowledge of the functions and the relationships of the structures which he saw in the plant he described [26, p. 160]. Again, use of simile was prominent in these sixteenth-century herbals to convey what the plant was like without measurements, without terminology to express shape and arrangement, and without botanical standards and specifications. While modern botanical descriptions are often incomprehensible to all but a few specialized botanists, even a cursory reading of the general descriptions of common plants, such as violets, potatoes, and roses, shows the need for a technical language that would convey a univocal meaning [26, pp. 161-162].

Contributions of Albert Durer to Plant Description

Improvements in illustrations, which occurred about 1530, can also be attributed to advances in the technology of book illustration, the use of copper engraved plates, and the pervasive influence of Leonardo da Vinci and Albert Durer, both of whom began to draw for wood engraving. The quality of their work was so impressive that it was impossible for botanists to remain satisfied with rudely conventional drawings of the early sixteenth century [26, p. 203]. Durer provided one of the earliest ecological drawings, whose detail far exceeds the capability of Renaissance botanical language to capture it. Durer's realistic drawing of "The Great Piece of Turf" (1503) is remarkable for its artistic precision: each plant is portrayed as an authentic representation, while all are brought together in a natural habitat to yield a superb but realistic work of art.

In botany, the result of rapid improvements in illustration resulting from art and typography unfortunately reduced, for a time, the power of the verbal description. In herbals with woodcuts, there was no difficulty in keeping a just balance between text and illustrations, as these were printed together. But in a book with metal engravings, which had to be printed separately from the type in the form of relatively expensive plates, the pictures easily became the primary feature of the work and lost their relationship to the text, which then declined in coordinated verbal/visual precision. Unlike medical and anatomical books, where visual illustration remained essential to verbal description for effective instruction, herbals in the seventeenth century became merely picture books: illustration became the master of description, rather than the servant of accurate verbal description [26, pp. 245-246]. Until the study of plants assumed a scientific footing and knowledge of plants gained a nomenclature to embody detail, botanical illustrations as art continued to usurp the place of precise description.

Leonardo da Vinci's Contribution to Mechanical Description

Tracking the development of technical description again requires recognition of da Vinci's work in mechanical description, which even more than his works on anatomy anticipates technical description and ultimately technical specifications as these are used in twentieth-century technical forms. Leonardo's notes and drawings of inventions, many of which he derived and then improved upon from earlier Italian inventors, anticipated mechanical drawings of the late seventeenth century. Da Vinci's contribution—multi-dimensional presentation—rests on his belief in the importance of perspective as controlled by geometry. For da Vinci, geometry meant not only proportions and measurement but also analysis of the nature of surface, line, and point. So it came about that the overwhelming majority of Leonardo's first speculations about science, and in particular the science of man, concentrated on light and shade, vision, perspective, proportion, and measurement. Through perspective he approached nature as a set of forms or geometrical figures. This geometry he applied not only to the surfaces of things which

the painter sees and paints but also to the forces, or powers of nature, which produce all action and movement [29, p. 21].

Da Vinci's mechanical drawings included engines of war, the helical screw helicopter, lens grinders, numerous drawings illustrating the study of perspective and the workings of physical laws, distillation equipment, designs for clocks, the universal joint, the hydraulic pump, waterwheels, and textile machines—just to name a few of his drawings. Like his plant descriptions, his mechanical drawings contain sketches of machines, with mechanical and mathematical applications. For example, when he sketched how light and heat worked when they concentrated in a pyramid, he drew the pyramid next to the following statement:

> The heat of the sun which is found on the surface of a concave mirror will be reflected in converging pyramidal rays to a single point. As often as the point itself, ab, or cd, enters the base, fg, so many times will the heat be more powerful than that on the surface of the mirror [quoted in 29, pp. 59, 88].

In many of his drawings, such as those of the textile shuttle, he merged process analysis with description. In a sense, Leonardo provided some of the early specification writing. The caption beneath the drawing of the shuttle is as follows:

> In this case it should be understood that when the two pipes h and k are together, as you see, joined, that it is the function of h to pull and of k to let go. But know that h would never pull away to the shuttle to itself if it were not locked itself in a, by means of the spring driven from m. But that m when the shuttle was given to h, did not touch the spring a; for the reasons that had it touched it, the spring would be released, and h would not be able to lock itself. But when h receives the shuttle it does not lock itself without a little motion under H and in this time K driving on its companion and accompanying it, moves outside n, and remains unlocked, and a in the same time locks itself and pulls its own shuttle to the whole motion [quoted in 30, p. 58].

The weakness in all Leonardo's mechanical drawings lay in the lack of precise engineering nomenclature to fully explain the concept he visualized, a deficiency which may be due partly to his lack of extensive training in mathematics.

Renaissance Books on Navigation

While navigation books would seem a likely source for the study of English Renaissance technical description, these books, many obviously written to be used as teaching tools, never contained the highly detailed verbal description coordinated with visual presentation that is found in anatomical texts. Although many navigation books contained descriptions of navigational instruments, many contained no visual illustrations, or the visual illustration was not effectively coordinated with the verbal description, as Chaucer had done in his *Treatise on the Astrolabe*. Writers apparently assumed a sophisticated readership having some familiarity with these basic instruments for charting voyages. Or, as Howes

suggested in *The Third Vniversity*, many of the books might have been written to supplement oral instruction in navigation taught in extension courses in London [7]. Writers assumed that students, who were aware of what the mechanisms looked like, needed more written process description than visual representation.

Because knowledge of the sea was so important to England's defense and commercial well-being, Howes mentioned the importance of the study of navigation. He specifically mentioned "a lecture of Cosmography read in the Blacke fryers in the house of Adrianns Marius" and "a cosmography college called Trinitie, housed in Stepford, founded by Henry VIII for the royal navy" [7, p. 981]. Unlike writers of anatomy books who knew that many readers would not have access to cadavers or public dissections, writers of navigation books written to instruct beginning seamen may have assumed that books would simply summarize instruction given during lectures where students had the opportunity to handle each device as it was discussed by the instructor. In short, the lack of detailed description of navigational mechanisms suggests the continuity of the oral tradition in navigational instruction. Many of these books present their instruction as a dialogue between two seamen, one who asks questions and one who responds—a method reminiscent of classroom disputation.

Barlowe's The nauigators supply [31]

Most navigation books described mechanisms and processes for using them. Like the herbals and even many medical books, these navigation books use a conversational, first-person narrative. For example, in *The nauigators supply* (1597) Barlowe used first person to describe the appearance, use, and care of five major navigational instruments [31]. Each instrument appears in a separate chapter, which includes a copperplate drawing for the most complex instruments. The description was not carefully integrated with the visual, and the device was not labeled to correspond strictly to its description. Barlowe apparently assumed that the reader had some experience with these instruments and did not need a presentation carefully correlated with the text. He may have assumed that the reader might have in his hand the devices described in the work.

The organization of some of the descriptions and instructions for using the instruments is similar to that used by Chaucer in describing the astrolabe. For example, the descriptions of the hemisphere and the pantometer follow this pattern (Chapters 4 through 8 are the how-to chapters):

Chap 1. The Pantometer. The Authors purpose in this Treatise
Chap 2. The Declaration of the partes of the Pantometer
Chap 3. The Mechanicall Description of those Partes
      [Parts are tabulated and appear on the page facing a fold-out drawing
      of the pantometer (see Figure 18).]
Chap 4. The exact finding of Altitudes or Heights
Chap 5. The obseruing of the Variation at Land by the Sunne
Chap 6. The obseruing of the Variation at Land by the Starres

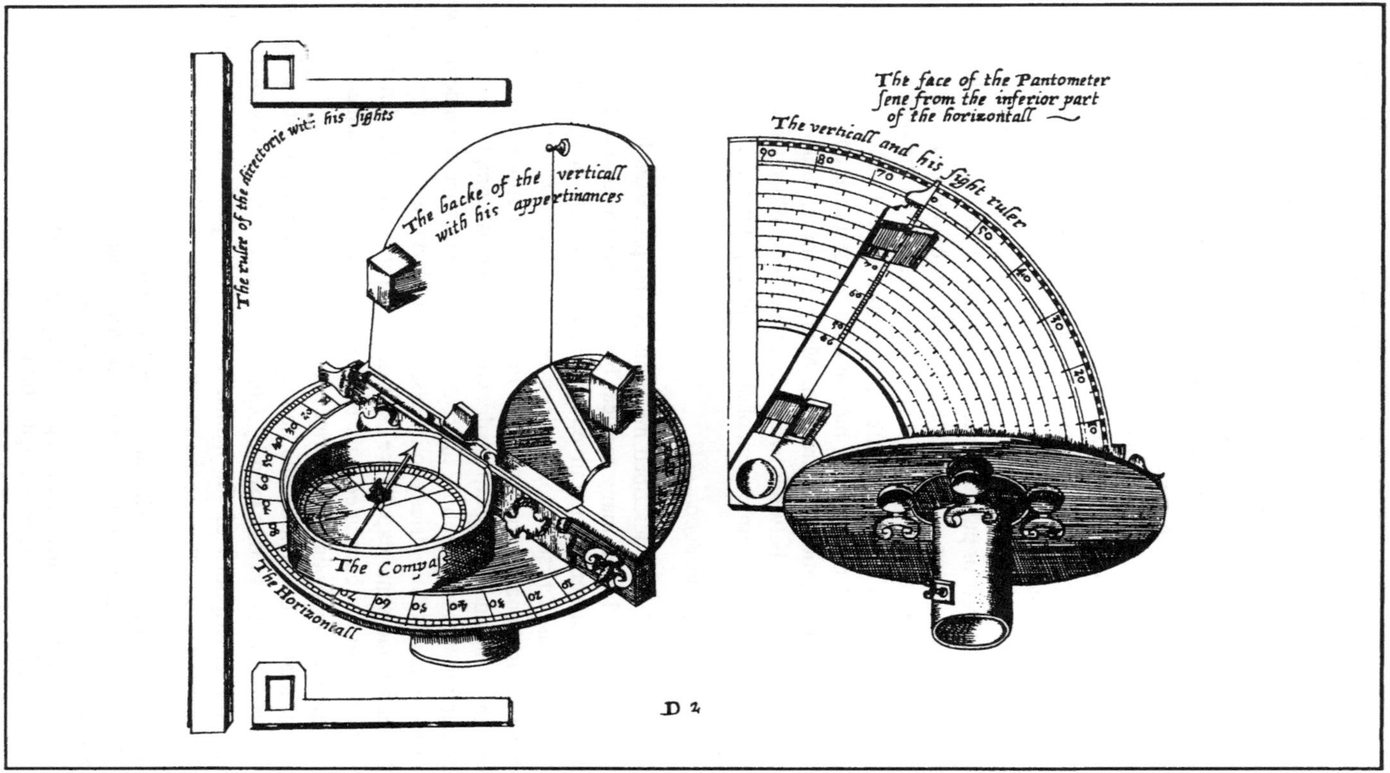

The face of the Pantometer sene from the inferior part of the horizontall

The vertical and his sight ruler

The backe of the verticall with his appertinances

The ruler of the Horizontall with his sights

The Compas

The Horizontall

D 2

Figure 18. The pantometer, from W. Barlowe, *The nauigators supply* (1597) [31].

Chap 7. The maner of obseruing the Variation at Sea
Chap 8. The taking of the height of the Storme at Sea [31].

Barlowe provided a general rather than a detailed description. His approach anticipates the method of technical description used today in parts and specification manuals: introduction, listing of parts, descriptions of parts, then use of the device.

In Chapter 2, describing the parts of the pantometer, Barlowe used a general spatial arrangement that emphasized partition, which he integrated with the drawing of the pantometer (Figure 18). He opened "The Declaration of the partes of the Pantometer" with the following paragraph:

> The chiefe parts of this Instrument are two: The *Horizontall*, to be placed alwayes equidistant to the plaine of the *Horizon*, which it representeth. And the *Verticall*, perpendicularly erected vpon the former, and therefore in power any *Azimuth* or *Verticall*, whereof also it is named. In the Horizontall there are two Semicircles; The one hath the ordinarie Points of the Compasse: the other the common degrees of a Circle, both of them meeting in one Diameter. The round voice space in the middle of the *Horizontall*, must be filled vp with the Centre-pinne thereof, in such sort, that being from vnderneath fastened thereinto, the one halfe and more of the said Pinne, somewhat beyond the Diameter of his thicknes, shalbe cut away, and be made euen with the vpper side of the *Horizontall*. But the other part remaining, shall stand directly vpright of a conuenient length aboue the *Horizontall*: And this I terme the *Axis* of the *Verticall*: because by meanse of it the *Verticall* is turned round about vpon the *Horizontall*, and made to take any maner of position [31, p. C4].

In the English Renaissance, navigation, like medicine (but unlike botany), was clearly a well-established field that possessed its own nomenclature. Books on navigation explained the application of geometry to charting voyages, mapping, and determining the location of ships. They also contained extensive tables of solar and lunar declinations and descriptions of coastlines with a drawing to show what various coastlines might look like from the deck of the ship as it sailed close enough for sailors to catch sight of a land mass. Unlike herbals, navigation books retained a balance of verbal and visual displays of information, although the quantity and quality of visual aids increased as sophistication in navigation and print technology also increased.

Books on Military Science

Howes's *The Third Vniversity* also mentioned the availability of extensive instruction and texts on military science [7]. These works, such as that by William Bourne, *The Arte of Shooting in Great Ordnaunce* (1578) [32], were designed as written summaries, rather than self-teaching texts, to accompany extensive oral instruction on artillery. In contrast, however, books on fortification contain detailed descriptions of fortifications intermixed with procedures for building

these structures. These books, like the books on artillery, were apparently used to supplement oral instruction, but the written discourse provided a record of details for actually building the fortification. Howes made very clear that men skilled in fortification were highly respected. Every gentleman was expected to have a command of the subject:

> Furthermore, here in this cittie are abiding very ingenious enginers, men very experienced in mining, sayling, and devising stratagems of war and men very skilfull in the art of pyrotechnic, or the fire workes, and of fortification: And all these be but members of the art military of Polemica, of which is the most noble and Heroicall Art, being the proper occupation and profession of Princes and gentlemen, . . . for it is the art which hath enobled all the greatest families in the world, and which protecteth and preserueth all other good arts and sciences in safety [7, p. 985].

That books on this topic would be hot commercial items is thus not difficult to believe.

The Development of Graphics

Steady improvement of graphics in technical books is particularly evident when we compare, in chronological order, books on military science from 1489 to 1639. This type of how-to book comprises a large category of Renaissance technical description. The *Short-Title Catalogue* records more technical writers who devoted their energies to military science than writers who focused on any other technical area besides medicine.

The first published English book on military science was Caxton's translation of Christine DePisan's *The Fayt of Armes & of Chyualrye* (1489), which contained no visual aids to support his instructions [33]. As we have seen in Chapter 3, the book is tightly organized, but ¶ marks are the only graphic cues that Caxton used to help the reader find his way through the text.

William Bourne's *The Arte of Shooting in Great Ordnaunce* [32] was one of the books referred to by Howes in *The Third Vniversity* [7]. Published in black letter, Bourne's book features basic woodcut drawings, integrated into the text, as shown in Figure 19. These drawings show the position in which the cannon barrel must be placed and include verbal instructions and illustrations on how to use plumb lines to ensure that the barrel is parallel to the ground. Tables showing rise to run—how the length of the barrel is affected by the increase in altitude above the horizontal—are poorly drawn with broken columnar lines and lack of clear column headings, as shown in Figure 20. This table also illustrates the lack of early tabular design expertise by Renaissance writers and printers. This work, when we compare it with seventeenth-century military works, becomes a useful text for tracking the development of tabular display of data.

Similar books using minimal graphic aids are Barnabe Rich's, *A Path-Way to Military practise* (1587) [34] and Flauius Renatus's *The Fovre books*

25   The Arte of shooting

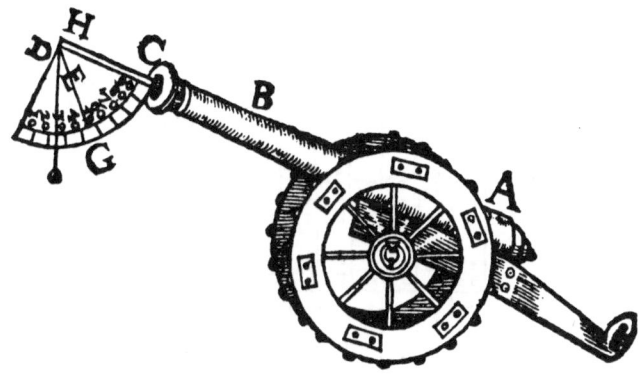

ſirſte take the Quadrant, and put the rule of the Qua-
drante E into the mouth of the peece C and then putting
vp oʒ downe the taple of the peece A, till the plummet
G fall vpon the coʒner of the Quadrant at D, then looke
whatſoeuer you ſee right with the vpper ſide of the Qua-
drante H, ſhall be leuell with the mouth of the peece,
and that is called the poynt blancke, foʒ that vppon a leuell grounde wythoute anye hylles, as vppon the ſea, that
all thinges ſtandeth ſo leuell, ſhall bee ryghte wythe the
Þoriʒon, that is to ſay, at the parting of the earthe and
the Skye, by the ſighte of youre eye : and then puttyng
downe the taple of the peece A, the plummet line G wyll
hange

Figure 19. "To dispart any peece of Ordnaunce," from William Bourne,
The Arte of Shooting in Great Ordnaunce (1578) [32].

Foote 10. long.	2	4	6	8	10	12	14	16	18	20	Inches.
	1/11	2/11	3/11	4/11	5/11	6/11	7/11	8/11	9/11	10/11	Partes.
Foote. 10. ½ long.	2	4	6	8	10	13	15	17	19	21	Inches.
	2/11	4/11	6/11	8/11	10/11	1/11	3/11	5/11	7/11	9/11	Partes.
Foote. 11. long.	2	4	6	9	11	13	15	18	20	22	Inches.
	3/11	6/11	9/11	1/11	4/11	7/11	10/11	2/11	5/11	8/11	Partes.
Foote. 11. ½ long.	2	4	7	9	12	14	16	19	21	24	Inches.
	9/22	9/11	5/22	7/11	1/22	5/11	19/22	3/11	15/22	1/11	Partes.
12.	2	5	7	10	12	15	17	20	22	25	Inches.
	1/2		1/2		1/2		1/2		1/2		Partes.
Foote. 12. ½ long.	2	5	7	10	13	15	18	21	23	26	Inches.
	7/11	3/11	10/11	6/11	2/11	9/11	5/11	1/11	8/11	4/11	Partes.
Foote. 13. long.	2	5	8	10	13	16	19	21	24	27	Inches.
	8/11	5/11	2/11	10/11	7/11	4/11	1/11	9/11	6/11	3/11	Partes.
Foote. 13. ½ long.	2	5	8	11	14	16	19	22	25	28	Inches.
	9/11	7/11	5/11	3/11	1/11	10/11	8/11	6/11	4/11	2/11	Partes.
Foote. 14. long.	2	5	8	11	14	17	20	23	26	29	Inches.
	10/11	9/11	8/11	7/11	6/11	5/11	4/11	3/11	2/11	1/11	Partes.
Foote. 14. ½ long.	3	6	9	12	15	18	21	24	27	30	Inches.
	1/30	1/15	1/10	2/15	1/6	1/5	7/30	4/15	9/30	1/3	Partes.
Foote. 15. long.	3	6	9	12	15	18	21	25	28	31	Inches.
	3/22	3/11	9/22	6/11	15/22	2/11	21/22	1/11	5/22	4/11	Partes.

Figure 20. Table showing mounting position for ordnance, from William Bourne, *The Arte of Shooting in Great Ordnaunce* (1578) [32].

This engine serueth for the xiiij. Chapter
of the fourth Booke. Folio. 55.

Figure 21. Drawing "Of the battering Ramme," from Flauius Renatus,
The Fovre books of . . . Martiall policye (1572) [35].

of . . . Martiall policye (1572) [35]. Renatus's work uses six full-page woodcut illustrations to show various methods of siege (see Figure 21), but all are placed at the end of the work with notation in the title of each woodcut of its relevance to the text.

By the turn of the century, military science books were longer, more detailed, and used textual/visual presentation for embodying complex military concepts. For example, Robert Barrett's *The Theorike and Practike of Moderne Warres* (1598) was published as a large folio with integrated drawings of the arrangements of regiments during battle, one large fold-out drawing of a military encampment, and twenty tables showing how to position soldiers, depending on the fronts shown by the enemy [36]. Use of fold-out drawings, with their intricately articulated verbal descriptions, shows the emphasis Renaissance military writers placed on detailed verbal/visual description.

Even more striking is John Bingham's *The Art of Embattailing an Army* (1629), which used copperplate illustrations, six fold-out illustrations of the arrangement of soldiers, all of which are situated at the appropriate point of the text and labeled with a title and the chapter to which they refer [37]. As in most seventeenth-century technical books, graphics and page design heighten the communicative quality of the text. Figure 22 shows the drawing and description of how the rhombus battle formation is to be used. As discussed in Chapter 3, Renaissance technical writers quickly discovered and used the art of fold-out pages to enhance the capability of complex graphic presentation.

Norwood's Fortification or Architecture Military [38]

Cooke's The Prospectiue Glasse of Warre, Shewing You a Glimpse of Warre Mystery, In Her Admirable Stratogems [39]

Looking more carefully at examples of technical description, we can examine Richard Norwood's *Fortification or Architecture Military* (1639), which can be classified as a technical description of a rampart [38]. In Chapter 2 Norwood provided a numerical list of "axiomes observed in fortification." The remaining chapters (3 through 12) provided instructional problems, apparently used during classes on military science, to teach students how to design ramparts. However, the technical descriptions of the ramparts would allow the material to be studied silently and then used in actual construction of a rampart. The verbal description of the same rampart, excerpted from Chapter 8 (see Figure 23, p. 224), illustrates the precision needed in constructing fortifications—a detail that clearly anticipates modern-day specification writing. Drawings were carefully integrated with the text:

> First then it is to be understood that that which you have drawne, as before we have shewed, namely the lines *KL. LO. ON. NF.* etc. is the outer edge of the Rampire, (as in this figure above) which Rampire may be in breadth

or thicknes inwardly 7. rods, or somewhat more or lesse as occasion requirs, for in a Fort of 12. sides or more, & of importance answerable, it may be 10. rods, and in a Fort of 4 bulworkes, being of lesse importance if it be 5. rods, it may be sufficient, and in small skonces much lesse, which thickenesse is here represented by *DV*. so that the line drawne by *V*. doth represent the inward side of the Rampire, being in the curtaine, flanke, and front, every where parallell or equidistant to the outside of the Rampire [38, pp. 84-85].

Other military works provided detailed technical descriptions of battle strategies. These books, such as *The Prospectiue Glasse of Warre,* . . . by Edward Cooke (1628) [39], contain verbal description supported with simple sketches composed of squares and rectangles to represent troops and artillery, as these would be arranged during battle. One of the characteristics of late Renaissance books on military science is their specificity, supported by visual presentation whatever the size and cost of the book. For example, the following excerpt describes "A Battell of 12000. foot and 4000. horse, with Ordnance in the midst, and on the wings." The verbal description is supported by a simple drawing of boxes to represent battalions, located on the facing page. Cooke tells the reader when to examine the drawing:

In the Battell are three battalions, containing 3000 men, (a thousand a peece) flankt with Muskettiers and with Muskettiers before them in the same fashion as the rest. In the Reareward (or left winge) of the battell, are likewise three battalions of 500. a peece, Embattailed as the rest, with Muskettiers before them, in the same forme as the other: Behinde these battalions (for seconds) are foure battalions of 500 a peece, standing against the Interuals of their opposite battalions; which Interuals are 200 foot broad. . . . These foure battalions haue shot before them as the former, which with the rest may march forth to skirmish with the Enemie; or stand still or second them vpon their retreate, before the Battels joyne after being in the Reare to giue vpon your Enemies flanks as the other. The rest of the battalions are in Front but twelue foote distance one from another, and at three foote order.

The diuisions of Muskettiers are allowed six foote, that they may the better fall through, hauing giuen fire.

In the Reare of all two battalions, of a thousand a peece, standing iust behinde the three battalions of the Battell, about a furlong of[f]. On the Flanks of these are 800, Horse, 400. in each flanke, oblique wise, the better to start forth and inuiron the Enemy. In like manner are the Horse martialed in the outward flanks of the rest, but in greater numbers. Peruse the Figure.

By the winges of these two battalions are two field Peeces ready turned and bent to the Reare, to discharge vpon the Enemy, if he should with Horse or Foote giue vpon that part, if not, then these field Peeces may bee with ease brought from thence to some other place to annoy him in other wayes [39, p. 24].

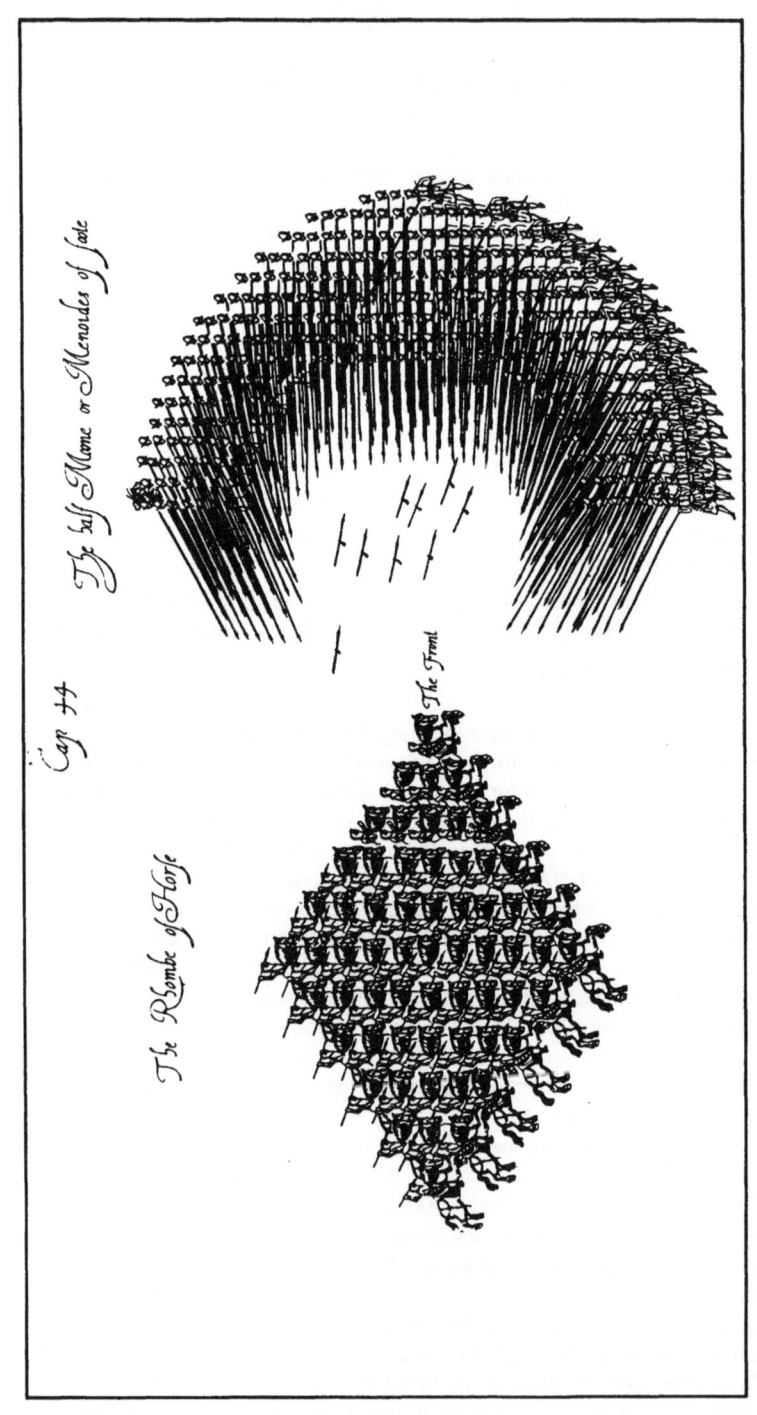

Figure 22. Drawing to accompany "Of the horse Rombe, and of the foote-half-moone to encounter it" (text on the facing page; in the original the illustration is oriented in the same direction as the text), from John Bingham, *The Art of Embattailing an Army* (1629) [37].

*Of the horse Rombe, and of the foot-halfe-moone
to encounter it.*

CHAP. XLIIII.

(1) He battaile framed in a forme of a Rombe, was first inuented by *Ileon the Thessalian*, and was called *Ile* after his name; and to this forme he exercised and accustomed his Thessalians. It is of good vse, becaufe it hath a leader on euery corner: in the front the Captaine, in the reare the Liuetenant, and on either fide the flank-commanders. (2) The foot battaile fitteft to encounter this, is the (3) Menoides or Creffent; hauing both the wings ftretched out, and within them the leaders, and being embowed in the middeft to enuiron and wrap in the horse-men in their giuing on: whereupon the horse-men ply the foot a farre off with flying weapons, after the manner of the Tarantines, feeking thereby to diffolue and diforder their circled frame of march. Tarentum is a City in Italy, the hofemen whetofore called Acroboliffs, becaufe in charging they firft caft little darts, and after come to handes with the enemy.

Daymen

NOTES.

1 *THe battaile in forme of a Rombe.*] Of the Rhombe is fufficiently fpoken in Chap. 6. before; and in the notes vpon the fame Chapter: The manner of framing of it, and the diuers kinds therof are there let down. The Theffalians

58 ## The Taktiks of Ælian, or

Words of direction for the Rhombe.

For the forming of the *Rhombes*, fee the 19 Chapter, and my Notes vpon that Chapter, §.6.

For the Creffent.

Firft order your body into a long fquare, Plagiophalanx.
1 The 2 file-leaders in the middeft of the fquare, ftand.
2 The next 2 on either hand, mooue forward one foot before the other two, their files mouing withall, and holding their diftance.
3 So the 4 next file-leaders each before other, on either fide a foot.
4 Then two more on either fide. aduance before the reft that mooued two foot a peece.
5 Then the 2 next on either fide, 3 foot apeece.

To reftore to the firft Pofture.

Face about. Moue all at once (excepting the 2 middle files) and take your firft ground.

Figure 22. (Cont'd.)

84

And so of others, But this by the way, now we returne from whence we have digressed.

CHAP. VIII.

Shewing how and in what forme, the Rampire, and Parapets are to be raised, and the Ditch to be sunke.

WE have shewed in the Chapter last before going, how to delineate the platforme of a Fort, and also how to stake out the same upon the ground, we will procede briefly to touch the rest.

First then it is to be understood that that which you have drawne, as before we have shewed, namely the

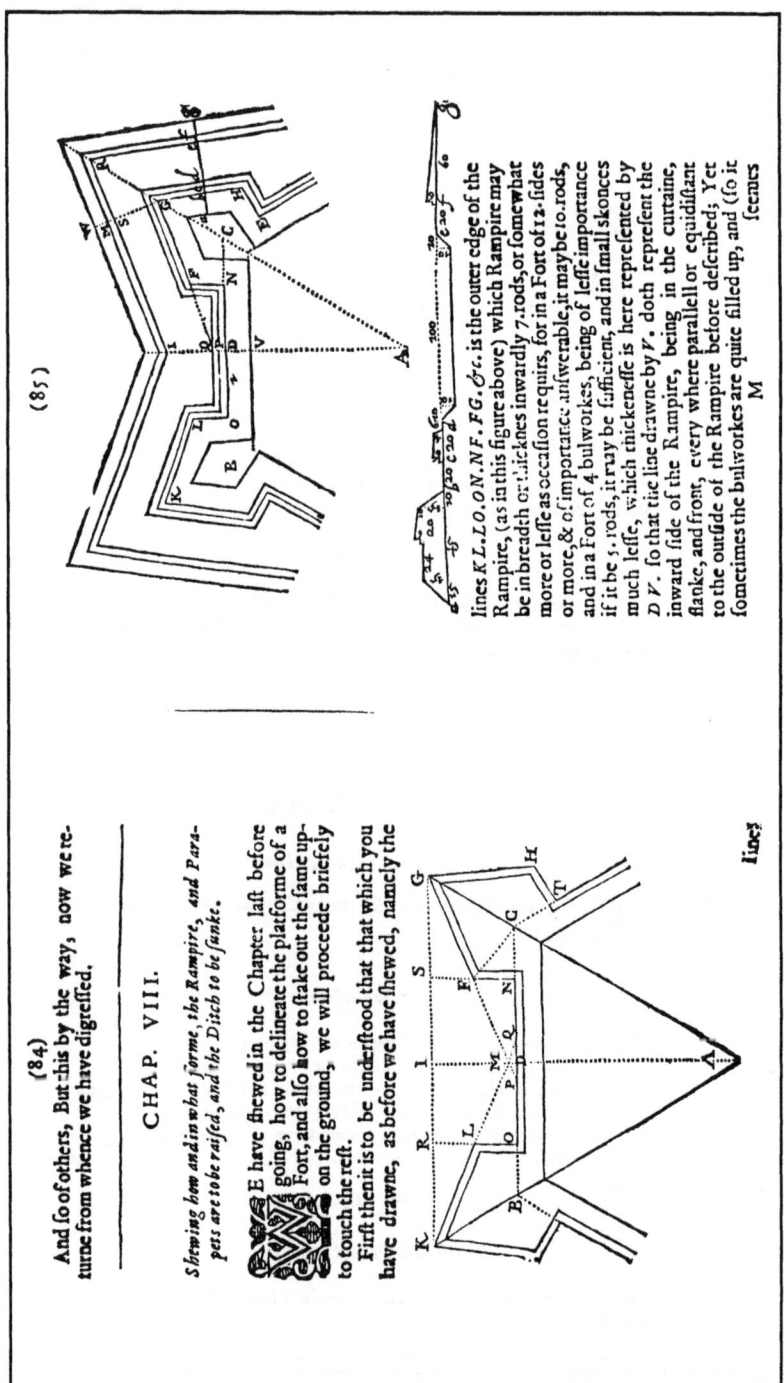

lines *K L. LO. ON. NF. FG. &c.* is the outer edge of the Rampire, (as in this figure above) which Rampire may be in breadth or thicknes inwardly 7.rods, or somewhat more or lesse as occasion requirs, for in a Fort of 12. sides or more, & of importance answerable, it may be 10.rods, and in a Fort of 4 bulworkes, being of lesse importance if it be 5. rods, it may be sufficient, and in small skonces much lesse, which thickenesse is here represented by *D V.* so that the line drawne by *V.* doth represent the inward side of the Rampire, being in the curtaine, flanke, and front, every where parallell or equidistant to the outside of the Rampire before described; Yet sometimes the bulworkes are quite filled up, and (so it seemes

M

Figure 23. Description of a rampart, from Richard Norwood, *Fortification or Architecture Military* (1639) [38].

The intense description throughout this book suggests that it was aimed at professional soldiers, committed to the art of war, who needed to study and interiorize these strategies for later use.

Blagrave's A Booke of the Making and Vse of a Staffe, Newly Inuented by the Author, Called the Familiar Staff [40]

The mathematical basis of military science suggests why numerous books were written on how to construct measuring devices to assist in fortification design. Iohn Blagrave's *A Booke of the Making and Vse of a Staffe* . . . (1590) includes in its subtitle the major readership of the work: "Newlie compiled; and at this time published for the speciall help of shooting in great Ordinance, and other millitarie seruices, and may as well be imployed by the ingenious, for measuring land" [40]. Chapters 1 and 2 explain the importance of the staff. Chapter 3 provides a technical description of the staff (see Figure 24). The remaining chapters provide instructions for making the staff. The entire work was designed to enable the reader to make the staff with only the written instructions available. The technical description, which is carefully detailed and integrated with the drawing of the staff, begins as follows:

> This staff consisteth of two seuerall partes, or rather two seuerall stavews, the one I call the standers staffe, the other the running staffe. Both in manner like, very plaine to conceaue: being each but as it were two streight rulers jointed together, to open and shutte even like an orinary paire of compasses.
>
> The standard staffe is here represented by this figure ABC, hauing his two legges or linnes AB and AC, ioynded at A, each legge consisting of 3. rules, or narrow boordes 5 foot in length, or breadth and thickness conuenient to carry it selfe streight, or some such wood as wil beste keepe from warping, gieweth and wrought together, so that botyh legs AB and AC being shutte close: ther may be within it a concanity of one ynch and an halfe square quite through to contained the running staffe for easy carriage. After this: for easier handling in use, let this standerd staffe be brought into an 8 square. And either lette the round or pomell of the ioinate A be of 4. inches diametre at the leaste, having a screwed pinne of mon through the centre with a scrwedbore on the under side to faste or lock fast the staffe being open at any angel or width [see the continuation of this description in Figure 24, left-hand page, line 2] [40, p. 38].

Blagrave's book was one of numerous books on military devices. William Bourne, a major writer of books on military science, devoted an entire book to *Inuentions or Deuises, Very Necessary for all Generalles and Captaines, or Leaders of men, as Wel by Sea as by Land* (1578) [41]. The book provides technical descriptions, simple drawings, and instructions for making mechanisms that could be quickly constructed, such as "The 54. Deuise" for determining "the goodnesse or the badnesse of [gun] powder" [41, p. 39].

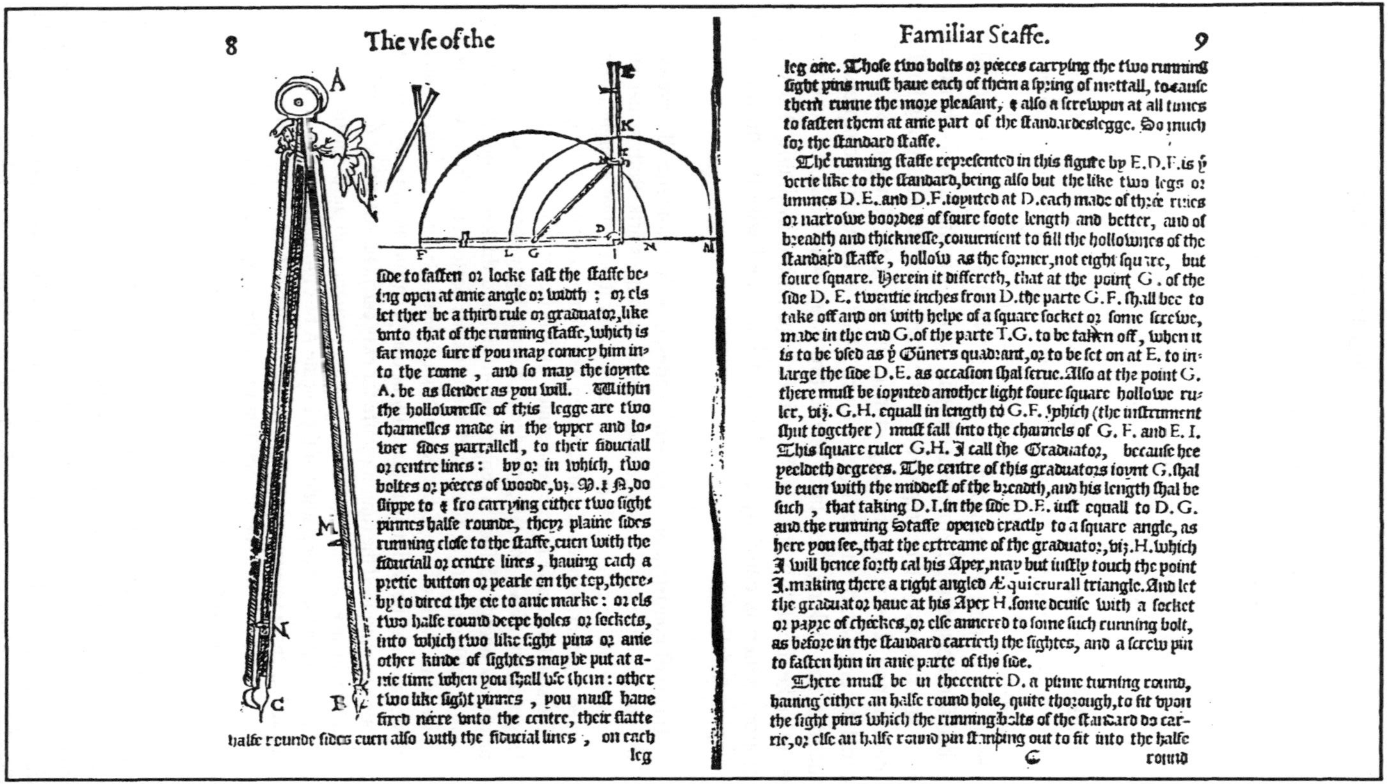

8 The vse of the

side to fasten or locke fast the staffe being open at anie angle or width ; or els let ther be a third rule or graduator, like unto that of the running staffe, which is far more sure if you may conuey him into the roome , and so may the ioynte A. be as slender as you will. Within the hollownesse of this legge are two channelles made in the upper and lower sides parrallell, to their fiduciall or centre lines : by or in which, two boltes or peeces of woode, viz. M. & N, do slippe to & fro carrying either two sight pinnes halfe rounde, theyr plaine sides running close to the staffe, euen with the fiduciall or centre lines, hauing each a pretie button or pearle on the top, thereby to direct the eie to anie marke : or els two halfe round deepe holes or sockets, into which two like sight pins or anie other kinde of sightes may be put at anie time when you shall use them : other two like sight pinnes , you must haue fixed neere unto the centre, their flatte halfe rounde sides euen also with the fiducial lines , on each leg

Familiar Staffe. 9

leg one. Those two bolts or peeces carrying the two running sight pins must haue each of them a spring of mettall, to cause them runne the more pleasant, & also a screwpin at all times to fasten them at anie part of the standardeslegge. So much for the standard staffe.

The running staffe represented in this figure by E.D.F. is y verie like to the standard, being also but the like two legs or limmes D.E. and D.F. ioynted at D. each made of three rules or narrowe boordes of foure foote length and better, and of breadth and thicknesse, conuenient to fill the hollownes of the standard staffe , hollow as the former, not eight square, but foure square. Herein it differeth, that at the point G. of the side D. E. twentie inches from D. the parte G. F. shall bee to take off and on with helpe of a square socket or some screwe, made in the end G. of the parte T.G. to be taken off , when it is to be used as y Guners quadrant, or to be set on at E. to inlarge the side D.E. as occasion shal serue. Also at the point G. there must be ioynted another light foure square hollowe ruler, viz. G.H. equall in length to G.F. which (the instrument shut together) must fall into the channels of G. F. and E. I. This square ruler G.H. I call the Graduator, because hee yeeldeth degrees. The centre of this graduators ioynt G. shal be euen with the middest of the breadth, and his length shal be such , that taking D.I. in the side D.F. iust equall to D.G. and the running Staffe opened exactly to a square angle, as here you see, that the extreame of the graduator, viz. H. which I will hence forth cal his Apex, may but iustly touch the point I. making there a right angled Æquicrurall triangle. And let the graduator haue at his Apex H. some deuise with a socket or payre of cheekes, or else annexed to some such running bolt, as before in the standard carrieth the sightes, and a screw pin to fasten him in anie parte of the side.

There must be in the centre D. a pinne turning round, hauing either an halfe round hole, quite thorough, to fit upon the sight pins which the running bolts of the standard do carrie, or else an halfe round pin standing out to fit into the halfe

C round

Figure 24. Description of the "staffe," from Iohn Blagrave, *A Booke of the Making and Vse of a Staffe . . .* (1590) [40].

Technical Description in How-To Books on Recreation

Recreation books that provided instructions on various sports provide additional examples of technical writing. Technical description, using varying degrees of detail, can be found, for example, in *Hungers prevention: or, the whole arte of fowling* by Gervase Markham (1621) and *A Booke of fishing with Hooke & Line* by Leonard Mascall (1590).

Markham's *Hungers prevention: or, the whole arte of fowling* [42]

As we have already noted in Chapters 2 and 4, Gervase Markham was perhaps the most prolific technical writer of the 1475-1640 period. Markham wrote approximately a dozen works on horsemanship, horseshoeing, estate management, archery, and fowling that enjoyed over four dozen editions. In his book on fowling, the first ten chapters provide general descriptions of methods of capturing fowl—the general method, kinds of fowl that are best captured, methods of taking fowl by "lyme-twigs," nets, and trained dogs. Chapter XI includes twelve pages describing the construction and operation of the fowling net for catching birds, which is then pictured at the end of the technical description/operational definition (see Figure 25). The following excerpt from Markham's description of the nets suggests his conversational style:

> First then for the Nets, you shall vnderstand that they are to be made either of very fine smale packthred, or else of very strong and bigge Housewifes thred, the mash small, and not abouve half an inch square each way, and the knots squarely knot without slipping, the length would be three sadome (or little lesse,) and the debth or bredth, would be one sadome and no more, it carryeth the fashion of the Crow-Nette, and must be bredged about (after the same manner,) with very strong small Coard, and the two ends extended vupon two small long poales, (sutable to the bredth of the Nette) in such manner as hath beene shewed before) [42, p. 113].

While Markham divided his book into chapters and used marginal headings to announce major sections, he penned his descriptions and instructions for use in undifferentiated prose. The type is large (12 point) and uncrowded, but careful reading is required to follow his description of the fowling devices and the means of using them. Thus, the drawing of the fowling apparatus becomes almost essential to the reader's understanding of the mechanism and its operation. The detail Markham uses throughout the book suggests that he expected readers to be able to follow his instructions and to build these "fowling devices" with little or no oral instruction.

Mascall's *A Booke of fishing with Hooke & Line* [43]

Part I of this book by Leonard Mascall provides a process description of how to catch various types of fish—salmon, trout, carp, bream, perch; the kinds of flies and baits to use; how to make hooks and sinkers; how to preserve fish; and how to

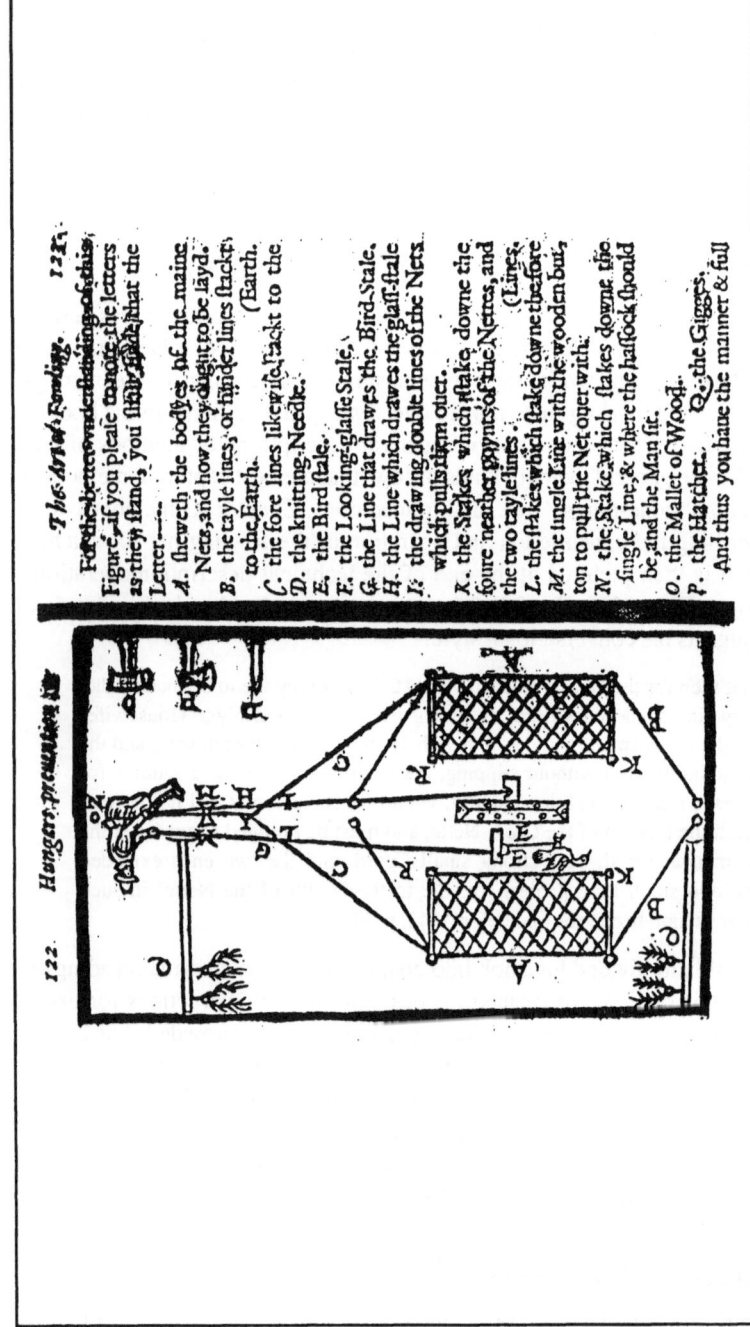

For the better vnderstanding of this Figure, if you please to marke the letters as they stand, you shall finde that the Letter——

A. sheweth the bodyes of the maine Nets, and how they ought to be layd.

B. the tayle lines, or vnder lines shackt, to the Earth.   (Earth.

C. the fore lines likewise shackt to the

D. the knitting-Needle.

E. the Bird stale.

F. the Looking-glasse Stale,

G. the Line that drawes the Bird-Stale.

H. the Line which drawes the glasse-stale

I. the drawing double lines of the Nets which pulls them ouer.

K. the Stakes which stake downe the foure weather poynts of the Nettes, and the two tayle lines.   (Lines,

L. the stakes which stake downe the before

M. the single line with the wooden but, ton to pull the Net ouer with.

N. the Stake which stakes downe the single Line, & where the hassock should be, and the Man sit.

O. the Mallet of Wood.

P. the Hatchet.   Q. the Giggs.

And thus you haue the manner & full

Figure 25. Description of a fowling net, from Gervase Markham, *Hungers prevention:*
or, the whole arte of fowling (1619 edition) [42].

stock and maintain fish ponds. Part II, *A Booke of Engines and traps to take Polcats, Buzardes, Rattes, Mice and all other kindes of Vermine and beasts whatsoever*, provides general descriptions with woodcut drawings of various traps. For example, Figure 26 shows "The bow trappe for Rats or other Vermine." The parts are labeled and described in the margin. The text beneath the drawings of the bow reads as follows:

> This engine or trappe with the bow, is made like a bore, of a whole peece of wood, with the lidde opening and shutting aboue, and this lide is unbent, shewing the left side and the lidde, and clicket, taken out: with holes and the string for to set him, as more plainly shall be shewed bent and set [43, p. 69].

On the following pages we read:

> This side shewes him bent, with the holes and pinnes how to bend him: as the pinne on the lidde is to holde the string bent. And also the pinne set aboue the clicket, is to order the string coming from the hole of the bridge unto the clicket, which must stay up the bridge crost and bayted, when he is set. The pikes are set to holde fast when he is put down [43, pp. 70-71].

The devices described in the book are relatively simple, and the quality of the woodcut drawings and the verbal description, which is fairly detailed, would enable readers to build the "engines" Mascall discusses.

RENAISSANCE GRAPHICS AND
THEIR LEGACY

Tracing the advances of technical description illustrates the interrelationship of the advancement of print technology, the advancement of knowledge, the sustained use of plain style as it persisted in Renaissance technical discourse, and the development of art as it was nurtured by both technology and knowledge.

First, print technology made possible the dissemination of information to a growing population of readers. An expanding readership meant that many kinds of knowledge could no longer be conveyed only by oral means. Instead, readers needed books that contained what they needed to know to perform tasks, apart from oral explanation.

Second, the growth of knowledge in quantity as well as complexity meant that texts, rather than oral transmission, were necessary to contain this new knowledge—organize it, then present it in ways that would preserve the accuracy of this increasingly complex information. The large number of books using technical description suggests that integrated verbal and visual descriptions were found to be the best means of organizing detailed information and conveying it to readers who needed to rely on silent reading to share in this changing world of new ideas. The need for visualizing technical processes, and hence the limitation of woodcuts, is evident when we compare the anatomical drawings of Roesslin,

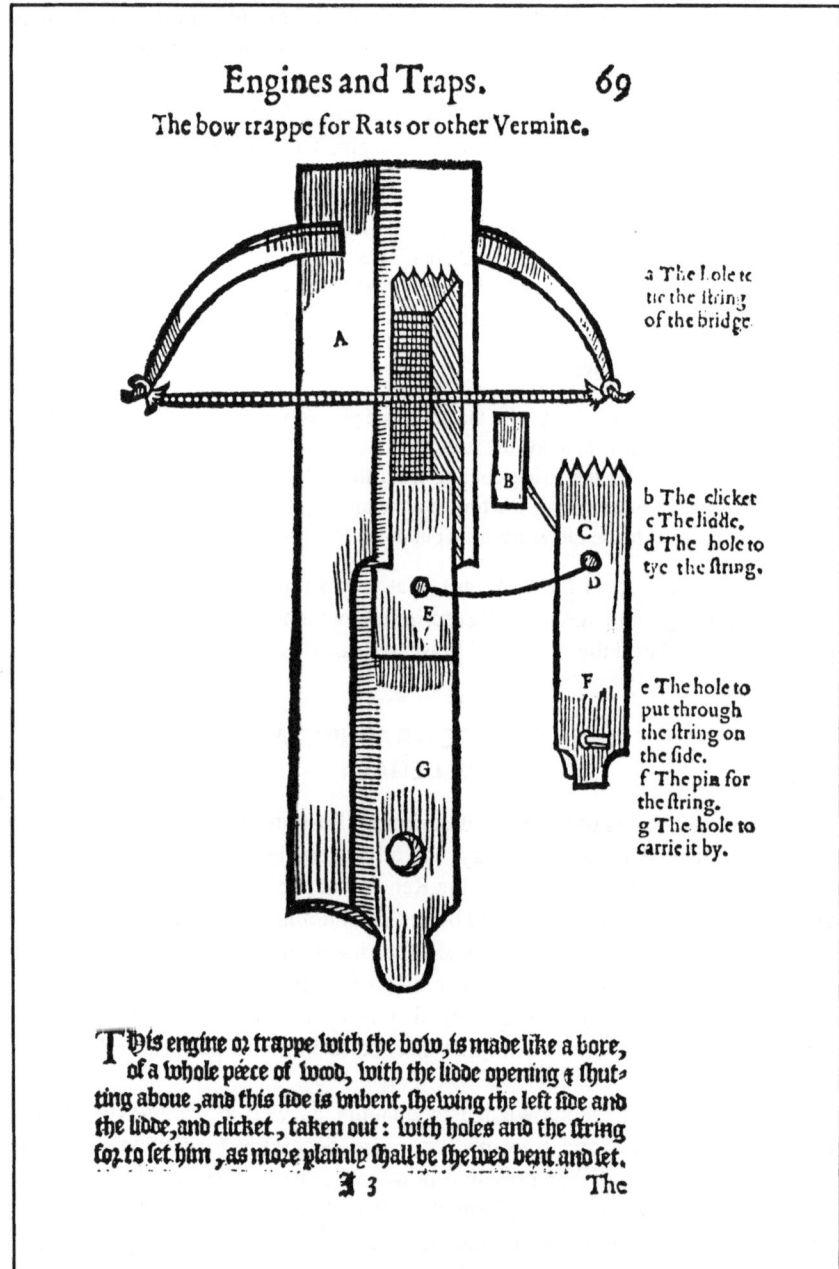

Figure 26. "The bow trappe for Rats or other Vermine,"
from Leonard Mascall, *A Booke of fishing with Hooke & Line* (1590) [43].

¶ Here followe certayne wayes of plan-
ting and grafing, with other necessaries herein
meete to be knowen, translated out of Dutch by L. M.

¶ To graffe one Vine on another.

YOu that wyll graffe one Uine vpon another,
ye shall (in Ianuary) cleaue the head of the
Uine, as ye doe other stockes, and then put in
your Uine graffe or syon, but first ye must
pare him thinne, ere ye set him in the head, then clap and
mosse him as the other.

¶ Chosen dayes to graffe in, and to choose
your syons.

ALso when so euer that ye wyll graffe, the best chosen
tymes is on the last day before the chaunge, and also in
the chaunge, & on the seconde daye after the chaunge, if ye
graffe (as some say) on the thirde, fourth & fift day after the
chaunge,

chaunge, it will be so many peres ere those trees bring forth
fruit. Which thing ye may beleue if ye will, but I will not.
For some do holde opinion, that it is good grafting from the
chaunge vnto the xviij. day therof, which I think to be good
in all the increase of the Mone, but the sower the better.

¶ To gather your syons.

ALso such syons or graffes which ye do get on other trees
the yong trees of thic or foure yeares, or ij. or. iij. yeares are best to haue graffs. Take them of no vnder bo-
wes, but in the top vpon the East side, if ye can, and of the
fayrest and greatest. ye shall cut them, ij. inches long of the
olde wood, beneth the syont. And when soeuer ye will graffe,
cut or pare your graffes taperwise from the syont two yn-
ches or more of length, which ye shall sette into the stocke :
and before ye set it in, ye shall open your stocke with a
wedge of yron, or harde wood, fayre and softly : then if the
sides of your clefts be ragged, ye shall pare them with the
poynt of a sharpe knife on both sides within and about, then
set in your graffes close on the out sides and also aboue : but
let your stocke be as little white open as ye can, and when
your graffes be well set in, plucke forth your wedge : and
if your stocke doe pinch your graffes much, then ye must
put in a wedge of the same woode for, to helpe your graffes : Then ye shall lay a thicke barke or pill ouer the cleft
from the one graft to the other, to keepe out the claye and
rayne, and so clap them two fingers thicke rounde aboute
the clifts, and then lay on Mosse, but Woll is better
next to your clay, or else to temper your clay with Woll
or butter, for it shall make it lyze closer and also stronger on
the stocke heade. Some take Woll next the clay and to wap-
peth it all ouer with linnen clowtes : for the Woll being
once moyste, will keepe the clay so a long tyme. And other
some doe take Woollen clowtes that haue bene layd in the
lyees of Woadewood, or such lyke bitter thing, to keepe
creeping Wormes from comming, vnder to the Graffes.

ije

of a Hoppegarden 23

If you haue softe greene Rushes abrode in the deawe and the Sunne, within two or thre dayes, they will be lythe, tough, and handsome for this purpose of tying, which may not be forgotten, for it is most certaine that the Hoppe that lyeth long vpon the grounde before be be tyed to the Poale, prospereth nothing so well as it, which sooner attayneth therevnto.

¶ Of hylling and hylles.

Noe you must beginn to make your hils, and for the better doing thereof, you must prepare a tole of Coopers Addes, but not so much boowing, and therfore lykest to the netherpart of a trowell, the poole whereof must be made with a round hole to receyue a helue, lyke to the helue of a Mattocke, and in the poole also a naple hole must be made, to fasten it to the helue.

This tole is here about better proportioned, than in that on the other side following.

This

A perfite platforme 22

one ende of the roune woulde be full before the other, whereas notoe they shall lye euen and sharpe aboue, lyke an Hayestacke, or the ridge of an house, and sufficiently defende themselues from the weather.

If you thinke that you haue not Poales ynowe to tyle the roune, pull downe the wydths or bandes lower, & your roune will be lesse.

¶ Of tying of Hoppes to the Poales.

When your Hoppes are growne about one or two foote high, bynde vp (with a Rushe or a Grasse) such as declyne from the Poales, twynding them as often about the same Poales as you can, and directing them alwayes according to the course of the Sunne, but doe it not in the meaning when the deaw remayneth vpon them, if your leysure may serue to doe it at any other time of the day.

This figure is more than halfe filled with Poales, and not filled vp to the top.

If

Figure 28. "Of the presurtion of Poales," from Reginald Scot, *A Perfite platforme of a Hoppe Garden* (1574) [45].

Vesalius, and da Vinci with the rude woodcuts, such as those found in farming and gardening books. Figure 27, from Mascall's *A booke of the art and maner, how to plant and graffe all sortes of trees* [44], and Figure 28, from Reginald Scot's book describing how to raise and harvest hops, *A Perfite platforme of a Hoppe Garden* [45], show how inadequate woodcuts were at capturing detailed processes and why copper-engraved plates would, by the close of the Renaissance, become the norm for graphic presentation.

Because improved social, professional, and economic status came to depend on education, readers demanded—and booksellers saw the importance of providing—books that met the interest, comprehension, and learning needs of a variety of readers. At the same time, the increasing availability of texts with better visual and verbal presentation encouraged the increasing numbers of English readers thirsty for works written in the vernacular. The demand for these books made the development of informative, reader-oriented texts essential to book printers who knew their own existence depended on the sale of successful books.

Third, improvements in print technology gave impetus to improvements in art that emerged from the Renaissance. Copper engravings could capture and convey details that had been unavailable to medieval illustrators. But the emergence of new knowledge in fields such as human anatomy clearly nurtured artists such as da Vinci to develop artistic means of expressing the new approach to anatomy. Enhanced powers in the capabilities of visual representation likewise nurtured the need for more precise verbal description and ultimately the development of technical nomenclature in fields besides medicine, military science, and naval science, which had been culturally entrenched for centuries.

The development of technical description, particularly the correlation of verbal and visual, the merging of page design with graphic presentation, the placement of graphics within the text at the appropriate place, the increasing emphasis on graphics which could capture detail as well as context—all created a method of using graphics and text that prevails today. For that reason, many of the graphics examples discussed in this chapter use techniques that are readily familiar in modern technical writing.

In short, technical description developed as knowledge and the products of knowledge developed. The precision of technical description becomes a means of demarcating which fields continued to rely on primarily oral transmission and which moved toward textual transmission: general descriptions suggest that readers could expect some oral instruction reinforcement; detailed descriptions suggest that readers were expected to learn the information with little oral supplement. While orality had supported general art and limited the need for verbal expression, textuality was characterized by the growing complexity of presentation of knowledge. Technical description emerged as one manifestation of the Renaissance commitment to knowledge.

REFERENCES

1. W. J. Ong, *Ramus, Method, and the Decay of Dialogue*, Harvard University Press, Cambridge, Massachusetts, 1958.
2. K. O. O'Keeffe, *Visible Song: Transitional Literacy in Old English Verse*, Cambridge University Press, Cambridge, England, 1990.
3. M. T. Clanchy, *From Memory to Written Record: England, 1066-1307*, Harvard University Press, Cambridge, Massachusetts, 1979.
4. H. S. Bennett, *English Books & Readers 1475-1557*, Cambridge University Press, Cambridge, England, 1952.
5. W. J. Ong, *Orality and Literacy: The Technologizing of the Word*, Methuen, London, 1982.
6. L. B. Wright, *Middle-Class Culture in Elizabethan England*, Cornell University Press, Ithaca, New York, 1958.
7. E. Howes, The Third Vniversity of England, Or a Treatise of the Fovndations of All the Colledges, Ancient Schooles of Priviledge and of Hovses of Learning, and Liberall Arts, Within and About the Most Famovs Cittie of London, in *The Annales or General Chronicle of England*, I. Stow, London, 1615 [STC 23338].
8. P. M. Jones, *Medieval Medical Miniatures*, University of Texas Press, Austin, Texas, 1985.
9. R. Herrlinger, *History of Medical Illustration, from Antiquity to 1600*, Editions Medicina Rara, New York, 1970.
10. W. Hammond, *The Method of Curing Wounds Made by Gun-shot, Also by Arrowes and Darts, with their Accidents*, London, 1617 [STC 19191].
11. R. T. Gunter (trans.), *Chaucer and Messahala on the Astrolabe*, Vol. 5 of *Early Science at Oxford*, Eight Volumes, Scribners, Oxford, 1919.
12. G. Chaucer, *The Riverside Chaucer* (3rd Edition), L. D Benson (ed.), Houghton Mifflin, New York, 1987.
13. C. O'Malley and C. M. Saunders (trans. and eds.), *Leonardo da Vinci on the Human Body*, Schuman, New York, 1952.
14. E. Belt, *Leonardo the Anatomist*, Greenwood, New York, 1955.
15. A. Vesalius, *De Humani Corporis Fabrica, Libri Septem*, Basileae, ex officina Joannis Oporini, Anno salutis reparatae 1543, Mense Junio.
16. T. Gemini, *Compendiosa Totius Anatomie Delineatlo Aere Exarata*, N. Udall (trans.), London, 1553 [STC 11716].
17. C. Singer (trans. and ed.), *Vesalius on the Human Brain*, Vol. 4 of *The Wellcome Historical Medical Museum*, E. Ashworth Underwood (ed.), Four Volumes, Oxford University Press, London, 1952.
18. R. L. Lind (trans.), *The Epitome of Andreas Vesalius*, M.I.T. Press, Cambridge, Massachusetts, 1949.
19. A. Read, *A Description of the Body of Man*, London, 1634 [STC 20783].
20. E. Roesslin, *The Birth of Mankinde, Otherwyse Named the Womans Booke*, T. Raynald (trans.), London, 1598 [STC 21160].
21. H. Von Braunschweig, *The Noble Experyence of the Vertuous Handy Warke of Surgeri*, London, 1525 [STC 13434].

22. A. Paré, *An Explanation of the Fashion and Vse of Three and Fifty Instruments of Chirvrgery*, H.C. (trans.), London, 1634 [STC 19190].

23. W. A. Emboden, *Leonardo DaVinci on Plants and Gardens*, Discordes and The Armand Hammer Center for Leonardo studies at UCLA, Los Angeles, California, 1987.

24. H. Von Braunschweig, *The Vertuouse Boke of Distyllacyon of the Waters of All Maner of Herbes*, London, 1527 [STC 13435].

25. W. Turner, *The First and Second Partes of the Herbal of William Turner, Doctor of Physick Lately Ouersene Corrected and enlarged with the Thirde Parte Later Gathered and Newe Set Oute with the Names of the Herbes in Grece Latin English Duche Frenche and in the Apothecaries and Herbaries in Latin with the Properties Degrees and Natyurall Places of the Same*, London, 1568 [STC 24367].

26. A. Arber, *Herbals: Their Origin and Evolution. A Chapter in the History of Botany 1470-1670*, Cambridge University Press, Cambridge, England 1986.

27. J. Gerarde, *The Herball or Generall Historie of Plants*, Two Volumes, London, 1597. Reprinted in *The English Experience*, Numbers 660A and 660B, Theatrum Orbis Terrarum, Amsterdam and New York [STC 11750].

28. R. King, *Botanical Illustration*, Potter, New York, 1979.

29. K. D. Keele, *Leonardo Da Vinci's Elements of the Science of Man*, Academic Press, New York, 1983.

30. K. G. Ponting, *Leonardo Da Vinci; Drawings of Textile Machines*, Moonraker Press, Humanities Press, New Jersey, 1979.

31. W. Barlowe, *The nauigators supply*, London, 1597 [STC 1445].

32. W. Bourne, *The Arte of Shooting in Great Ordnaunce*, London, 1578. Reprinted in *The English Experience*, Number 117, Theatrum Orbis Terrarum, Amsterdam and New York, 1969 [STC 3420].

33. C. DePisan, *The Fayt of Armes & of Chyualrye*, Westminster, 1489 [STC 7269].

34. B. Rich, *A Path-Way to Military practise*, London, 1587 [STC 20995].

35. F. V. Renatus, *The Fovre books of Flauius Vegetius Renatus, briefelye contaynionge a plaine forms, and perfect knowlede of Martiall policye*, London, 1572 [STC 24631].

36. R. Barrett, *The Theorike and Practike of Moderne Warres, Discourses in Dialogue wise*, London, 1598. Reprinted in *The English Experience*, Number 155, Theatrum Orbis Terrarum, Amsterdam and New York, 1969 [STC 1500].

37. J. Bingham, *The Art of Embattailing an Army*, London, 1629. Reprinted in *The English Experience*, Number 70, Theatrum Orbis Terrarum, Amsterdam and New York, 1968 [STC 162].

38. R. Norwood, *Fortification or Architecture Military*, London, 1639 [STC 18690].

39. E. Cooke, *The Prospectiue Glasse of Warre, Shewing You a Glimpse of Warre Mystery, In Her Admirable Stratogems*, London, 1628 [STC 5670].

40. I. Blagrave, *A Booke of the Making and Vse of a Staffe, Newly Inuented by the Author, Called the Familiar Staff*, London, 1590 [STC 3118].

41. W. Bourne, *Inuentions or Deuises, Very Necessary for all Generalles and Captaines, or Leaders of men, as Wel by Sea as by Land*, London, 1578 [STC 2431].

42. G. Markham, *Hungers prevention: or, the whole arte of fowling*, London, 1621 [STC 17362].
43. L. Mascall, *A Booke of fishing with Hooke & Line*, London, 1590 [STC 17572].
44. L. Mascall, *A booke of the arte and maner, how to plant and graffe all sortes of trees*, London, 1572 [STC 17574].
45. R. Scot, *A Perfite platforme of a Hoppe Garden*, London, 1574 [STC 21865].

CHAPTER 7

The Legacy of English Renaissance Technical Writing: New Perspectives on Basic Rhetorical Issues

If this book has been successful, then it has shown in a convincing way that what technical communication teachers and practitioners call "technical writing" existed in the Renaissance, and it has shown what subject matter it covered, what technologies were important to Renaissance readers, and what this technical writing looked like as we inevitably compare it with modern kinds of technical writing. Like any description of a genre of writing, my examination of examples of technical writing has been confined to a limited number of works, many of which I have examined from several basic perspectives: format and page design, audience, style, and integration of visual/verbal description. As the *Short-Title Catalogue* shows, many other examples of technical writing exist that are worthy of examination under the same rubrics in addition to others. These works will provide their own unique insights into the evolution of technical writing as an indicator of life, work, technology, and literacy in the English Renaissance.

The extensive use of example pages from the works discussed is deliberate: as we examine early technical writing, we need to know that the visual aspect of discourse was as important to Renaissance technical writers as it is to modern technical communicators. Literacy did not eliminate the importance of images in favor of words, at least in technical communication. Print simply created opportunity for a richer mixture of image and word, format and concept, graphics and text, to capture meaning. Mnemonics remained important but not as the sole

means of instruction. Graphics, as visual mnemonics, aided readers who also had access to text, which, by the seventeenth century, contained more information than could be remembered by oral transmission.

Locating additional works for examination will require a search of the *Short-Title Catalogue* and other collections of early English printed books, pamphlets, and broadsides, and classification of the works according to many of the major categories used in developing this study and the ten articles from which it evolved [1-10]. But these categories illustrate, at best, only one approach that can be used to explore these works.

Also needed are monograph-length studies of accounting books and epistolaries, as these were important Renaissance technologies, a fact suggested by the substantial number of works published on each topic during the sixteenth and early seventeenth centuries. Specific studies devoted to legal writing, policy writing, surveying, geography, and special fields of military science will show important developments in the discourse, in the technical expertise of these subject areas, as well as the political, religious, and social issues that defined the English Renaissance milieu. The numbers of books on agrarian and horticultural topics as well as navigational science suggest the importance of these topics to English readers. A close, chronological study of each category of works is needed to track the relationship between development of technology and development of style in conveying that technology to the target audience. Determining the areas that produced the most technical books can indicate what technologies were most important to English Renaissance readers and how these technologies affected life in the English Renaissance. Describing the characteristics of the discourse communities represented by each category of books would provide an illuminating study.

IMPLICATIONS OF RENAISSANCE
TECHNICAL WRITING

My broad discussion of a few works from this large body of technical instruction books reveals that many of the concerns that have been the focus of modern technical communication research were also concerns of Renaissance technical writers. Tracing the nature of format, audience awareness, style, and graphics in Renaissance technical writing—simply as topics for initiating historical exploration of technical writing—helps establish the history of technical writing while casting new perspectives on important issues in the history of rhetoric and language.

Clearly, anyone in technical communication cannot avoid looking at historical examples through the lenses of the present. And for that I make no apology. Even though I have presented Renaissance technical writing through the alembic of fundamental modern concerns in technical communication, I believe that we can see that literacy transformed technical writing; technology emerged from

knowledge; printing reshaped how work was perceived in medicine, agriculture, and military science; and ultimately, literacy and print nurtured each other. Evolution in technical books during the Renaissance supports these observations. Ong's argument that "literacy restructured consciousness" [11, pp. 126-127] is evident in technical writing. To extend Einstein's query [12, p. 8], the rapid changes in knowledge and print, seen through these books, suggests that printed technical books substantially changed human behavior and the ways in which people performed work as technology and literacy enabled that change. Experience moved from conglomerative statements, reminiscent of auditory transmission, to successive statements that developed an increasingly structured visual organization. Increases in the quantity of information that had to be captured in text favored the visual organization of knowledge and the use of perspective in the design of the printed page.

Purusal of the *Short-Title Catalogue* allows any interested Renaissance scholar to see that readers were receptive to a variety of books. While traditional English scholars have focused on selected works produced by Renaissance English writers, these scholars have unfortunately ignored the large body of how-to books that shed additional light on life and the evolution of language in England during the Renaissance, the characteristics of readers to which these books were targeted, and the literacy level of these readers. As is always the case in discussing technical writing, one cannot discuss a specific aspect of audience, style, or visual presentation without discussing or echoing all of them. For the consciousness of readers is inextricable from concerns about style, visual presentation, and the limitations and demands of printed text.

Audience

The profit motive of the Renaissance publishing business helps us understand 1) what works were successful and 2) that writers, who wanted to continue to write, were sensitive to the needs of their readers. We have only to examine books by Elyot, Markham, Lawson, Mascall, and Read, for example, to see how their awareness of the comprehension needs of their readers influenced their decisions about how to present their ideas. Examining and then comparing the most popular technical books, as suggested by numbers of extant copies, and books that were published in only one edition, may provide evidence (beyond what I have been able to discover) about the interests, reading levels, and needs of English Renaissance readers.

Similarly, the shift from poetry to prose exposition decreased the agonistic and eventually the hortatorical quality of technical writing. But the increase of knowledge and technologies, the need to sell books that adequately transmitted information to readers who were unknown to the writer, played a major role in the shift toward more specific, less didactic presentation. Thus, the emphasis on conveying information in print probably played a major role in objectifying

written discourse and removing much of its poetic and then oral quality. Increased technology to be communicated via written text necessarily placed increasing demands on writing, which took on new syntactical and structural characteristics to carry this heavier semantic load. Many writers also saw the value of visual aids to enhance the univocal meaning of their discourse. Tracing the shifts from orality to textuality and to visual presentation in technical writing—in terms of the needs of readers and the growing knowledge base—is clearly the subject for an intensive study.

Studying Renaissance technical books, particularly in terms of the characteristics and changes within the discourse communities of emerging disciplines and social groups, can fulfill Elizabeth Einstein's call for an assessment of the impact of printing on human consciousness, assessing "how printing affected ways of learning, thinking, and perceiving," "how access to printed materials affects human behavior" [12]. Undoubtedly, consideration of the intended audiences for these technical books will fill a major gap in our understanding of the power of print and knowledge on readers.

Format

In a major sense, technical writing represents the first concerted effort to control meaning by controlling typographic space. Because technical writing was audience specific, writers recognized that the spread of knowledge required accurate replication of concepts for readers. Knowledge, in order to grow, had to be able to have meaning in non-oral textual forms that could be preserved and reproduced graphically for rapid access. Renaissance technical writing enforces our belief that knowledge and its paradigms can exist only in print, because knowledge is textual and does not exist outside the word. Thus, technical writing attempted to transmit information onto the page by structure—organization of material and, increasingly, page design. Renaissance technical writers obviously believed in the possibility of a univocal meaning. Furthermore, print and pictures could capture reality. Eventually pictures and visual display could capture order and a meaning that they assumed was univocal and true.

Organization in technical writing exhibited more demonstrative forms than in other kinds of writing, particularly after Ramist rhetoric affected page design that emphasized overviews, visual organization, partition and division, and headings that revealed this partition. By the closing decades of the Renaissance, organization had shifted from aggregative, additive, redundant prose to highly structured, analytical prose in which the human voice was strangely absent.

Style

A study of Renaissance English technical writing against the philosophical background of Ong's *Orality and Literacy* [11] provides additional insights into the general shift from orality to textuality. Unlike literature, technical writing shed

its oral residue by the middle decades of the seventeenth century, as more and more technical writing used tables and charts to replace written discourse. In fact, the increase of tables, charts, and drawings announces the shift from oral to textual transmission of knowledge. (How can one speak a table?) Thus, echoing Ong, Renaissance technical writing, even more than writing of the standard literary genres, allows us to trace the analytical development of language, the development of rudimentary classifications into "artistically sequential, classificatory, explanatory examination of phenomena impossible without reading and writing" [1, p. 8]. Technical writing, perhaps even more than literary, religious, or philosophical writing, allows us to assess the effects of print on thought and memory, the effects of visual aids and word arrangement on thought and memory.

Of great importance are the insights that English Renaissance technical writing provides into the development of modern English prose. Examination of technical style shows conclusively that what came to be known as "plain English" existed in popular technical books long before Bacon or the Royal Society and that plain style was as much a characteristic of Renaissance discourse as was the aureate style used predominantly in philosophical discourse. Thus, even a cursory study of technical style, such as I have attempted in Chapter 5, lays to rest many long-held views concerning the rise of modern prose style:

1. The belief that tropes are inappropriate for scientific discourse was not founded on Renaissance technical documents, which used metaphor pervasively though carefully to enable readers to visualize meaning in the absence of precise nomenclature. Like modern technical writing, Renaissance technical writing illustrates effective use of metaphor and analogy, particularly in herbals and medical books which textualized the description of plants and processes and medical conditions without the benefit of scientific nomenclature. Scientific writing of the late seventeenth century rejected abuse of figures as they were exemplified in aureate prose; but plain style, which persisted from the earliest extant English prose, continued to use figurative language as a means of building meaning and capturing reality. A study of figurative language in Renaissance technical books would add to our understanding of how writers perceived their readers, how writers believed readers made meaning with the figurative language chosen. Similarly, a careful study of oral residue in tropes and schemes, as these were chosen to affect recall, will further enhance knowledge of the literacy of readers, as perceived by technical writers.

2. Renaissance technical style often exhibited use of first-person narrative, which, contrary to much modern technical style, was apparently not thought to reduce the effectiveness of the work in explaining how to perform a task or process. In many technical books, we can sense that the writer is talking directly to the reader because of the use of direct address. Most Renaissance technical writing uses a conversational style, which also suggests the nature of conversational Renaissance English. However, books that were designed about a

question-and-answer format, a descendent style of the oral tradition, became less popular by the early decades of the seventeenth century. Reading, as Ong has suggested, restructured consciousness: the structure of written discourse required an organization different from the embedding of orality in word form. Tracking changes in organization of technical books thus becomes a further means of tracking the demise of orality in modern English prose.

Renaissance technical writers thus employed what Halloran and Bradford call the fundamental metaphor, a metaphor to think and live by [13, p. 187]. As we have seen, there are enough of these kinds of metaphors in Renaissance technical writing to trace the trend to this point in the history of technical writing. An examination of Renaissance technical writing also brings into question the assertion that the use of tropes disappeared in technical writing because of the growing emphasis on plain and simple prose. Renaissance technical writing used both—plain style and tropes—in addition to headings, lists, and other format techniques to produce transparency and readability of text.

As Halloran and Bradford also pointed out, plain style has its limits. Bacon did not assiduously practice what he preached, and much technical writing uses a comprehensible style formed by style and tropes that were conjoined to appeal to readers' apparent interest in aesthetic phrasing while making the message clear. Thus, the problem from the perspective of rhetorical history seems to be that late seventeenth-century scientists were objecting to obdurate profuseness of philosophical style rather than what Chambers called the native English plain style that existed in a variety of technical books (Chapter 5). Literary historians have ignored the fact that Renaissance technical writing was experimenting with visual rhetoric to replace auditory patterns preserved in much undifferentiated text. Thus, visual organization employed visual rhetoric, as opposed to auditory patterns, which appealed to oral rendition.

Ultimately, however, tropes and schemes are inherently oral-based to reflect thought and recall mnemonics, and their relationship to format and the use of space and typographical arrangement to emphasize visual metaphor and mnemonic is an issue worthy of study in its own right. Format aids visual memory, as tropes aided the aural memory. Another issue for Renaissance technical writing, as for modern technical writing, is the subtle relationship that exists between voice and text. Halloran and Bradford's comment applies equally to both Renaissance and modern technical writing:

> While technical and scientific writing is primarily visual, no writing can ever divorce itself entirely from the auditory mode. The ultimate radical of presentation for any text is the human voice, as our persistent use of the term audience—the ones who audit—suggests. An important area for investigation in the rhetoric of science and technology would thus be the relationship between visual and auditory modes in the processing of texts [13, p. 191].

Visual Presentation

Visual presentation as a focus for the study of Renaissance technical writing is perhaps its most crucial, defining characteristic. For that reason, I have deliberately used numerous example pages from technical books to help students of discourse history see what Renaissance technical writing looked like. Full-fledged technical description, exemplified by the coordination of visual presentation with verbal presentation, represents the first definite break with orality. As Ong has pointed out, print suggests that words are things more than writing ever did, so that the surface of the text has a meaning apart from the spoken quality of the text [11, p. 126]. Visual and verbal texts were most often written to be used by the individual reader apart from supplementary oral instruction. Thus, technical description interiorized understanding—that is, memory became less important than the ability to access text as reference. A study of technical description—its visual presentation as well as its use of metaphor and analogy—can help modern readers resurrect the context in which the action depicted initially occurred.

Technical description which integrated text, page design, and visual aids provides the crucial markers of the development of technical writing in the English Renaissance. In technical description, we see writers creating text that reflects the needs of their readers. In response to these readers, we can see the writers' selection of structures for written discourse and a host of design techniques to portray meaning. Many of the examples presented in Chapters 3, 4, and 5 show the difficulty of separating technical writing from visual presentation. Aided by improving print technology, Renaissance technical writers quickly learned the value of art as an aid to communicating information. As silent reading replaced oral transmission of knowledge, increasing knowledge which needed to be textualized for consumption by an increasing literate population could be accessed by visual organization schemes far more efficiently than by writing that attempted to retain auditory schemes only.

The power of technical description, which merged the power of the visual with the written, thus helped to capture exact observation which was itself important to the development of science. Visual replication and the printed word supported a positivist, windowpane theory of reality, an objectified reality that could be held and studied. Without this form of communication and the belief that knowledge could be captured in text, knowledge would not advance. The means of building knowledge, which had to emanate from printed texts because of the limitations of human memory, can be seen in the late Renaissance technical books such as those in medicine and surgery.

FURTHERING THE SEARCH FOR OUR PAST

Concluding, as I began, by referring to Michael Moran's statement that the history of technical writing has not been written, I want to suggest other research

on the history of technical writing that awaits exploration. Again, if any book is successful it should illuminate territory for further inquiry.

1. A definition of technical writing. Since the 1970s, with the initial surge of demand for technical writing courses, defining technical writing has been problematic. Examining technical writing as it existed and developed during the Renaissance shows that it indeed attempted to adapt technology to the user, to use Dobrin's phrase (see Chapter 1); that it provided written discourse (to supplement and then to replace oral discourse) that would provide self-help for individuals eager for knowledge; that the Renaissance world of work was not easily separated from daily living problems and tasks of English men and women; that technical writing was value laden while being written to help individuals perform important daily tasks in the Renaissance world of work.

In defining Renaissance technical writing as it differed from other forms of printed discourse, research will inevitably lead into exploration of technologies of different eras—what they were; the shift from oral instruction for learning technology to written instruction; additional estimates, from the numbers of books for various audiences, of the literacy of readers and the literacy level of these readers; technologies (such as sewing and spinning) that remained orally based.

2. Rhetoric of technical writing. Renaissance technical writing offers a unique opportunity to examine applications of classical as well as Ramist rhetoric in the design of books. While I have provided initial exploration of the use of Ramist thought in many forms of writing, technical and non-technical, a study of classical rhetoric, as it was used by Renaissance technical writers, would help define the rhetoric of Renaissance technical writing as a basis for comparing rhetoric of technical writing in succeeding eras.

3. Analysis of style in specific kinds of technical books. What differences in style occurred in technical books as knowledge increased, as literacy increased? In arguing against both Jones's and Croll's positions, I have of necessity had to avoid detailed discussions of style as it reflects changes in literacy (see Chapter 5). I leave it to more sophisticated students of style to track the finer points of technical style development in the accretions of modern English prose.

4. Renaissance technical writing as a precursor of modern technical writing. What do applications of current social construction theories and the concept of discourse communities tell us about how technical writers perceived their audiences and their purposes with these audiences? On this issue, I am particularly interested in how successful technical books—those that enjoyed many editions—"made" meaning. What were the values of different knowledge communities, such as medicine, navigation, military science, and agriculture? What ideologies are represented in graphics as well as in the written discourse of these technical books? What does Renaissance technical writing look like from a postmodern perspective? In short, a reading of Renaissance technical books from existing research in social perspectives should further illuminate Renaissance technical writing. To my mind, Renaissance technical writers, without question, had a

universalist view of language: the world was definitely "out there," and the problem was using language to describe "out there" to enhance knowing for newly literate readers seeking to broaden knowledge. Ultimately, these early English technical writers exemplify logical positivism in language: sententious tropes and schemes were replaced by sharp description, metaphors were selected that clarified rather than suggested, non-dramatic language attempted to duplicate the writer's reality for the reader.

5. Finding and noting changes in books about agriculture, geography, and medicine. This provides an exciting opportunity to examine changes in literacy among readers. Noting the intended readers of many types of books, such as cookbooks, and noting changes in style, diction, and content can provide a useful tool for assessing the literacy of middle-class readers for whom historical evidence is lacking.

6. Technical writing in later periods. The quantity of published writing increased dramatically following the English Civil Wars, with a major outpouring of books from 1641 to 1700. Finding and describing technical books in English during the latter half of the seventeenth century seems a logical next step in following the evolution of the concepts that I have explored here for the 1475-1640 period in England. The evolution of American technical writing is also in need of exploration, as is an illumination of English technical writing of the eighteenth century, the nineteenth century, and the Industrial Revolution in particular.

In closing, then, I hope I have challenged students of the history of instrumental or applied rhetoric. My intent has been to mark a new area of language study which can be fully explored by scholars who are much better equipped than I am. I will be satisfied if I have provided a beginning on which they can build. For technical writing was clearly a substantial form of printed discourse, and it existed in quantity, albeit in manuscripts, in preprint societies. As a substantial form of discourse, it shares many characteristics with writing that has come to be revered because it is defined as imaginative or literary. As students of language and culture, if we are to understand the discourse of any period, we must explore not only literary and religious discourse, but pragmatic discourse as well.

REFERENCES

1. E. Tebeaux, Books of Secrets—Authors and Their Perception of Audience in Procedure Writing of the English Renaissance, *Issues in Writing, 3*, pp. 41-67, 1990.
2. E. Tebeaux, Visual Language: Format and Page Design in English Renaissance Technical Writing, *Journal of Business and Technical Communication, 5*, pp. 246-274, 1991.
3. E. Tebeaux, Ramus, Visual Rhetoric, and the Emergence of Page Design in Medical Writing of the English Renaissance: Tracking the Evolution of Readable Documents, *Written Communication, 8*, pp. 411-443, 1991.

4. E. Tebeaux, From Orality to Textuality: The Evolution of Technical Description in English Renaissance Technical Writing, *Issues in Writing, 3*, pp. 59-109, 1991.

5. E. Tebeaux, Renaissance Epistolography and the Origins of Business Communication, 1568-1640: Implications for Modern Pedagogy, *Journal of Business and Technical Communication, 6*, pp. 75-98, 1992.

6. E. Tebeaux and J. M. Killingsworth, Expanding and Redirecting Historical Research in Technical Writing: In Search of Our Past, *Technical Communication Quarterly, 1*:2, pp. 5-32, 1992.

7. E. Tebeaux and M. M. Lay, Images of Women in Technical Books from the English Renaissance, *IEEE Transactions on Professional Communication, 35*:4, pp. 196-207, 1992.

8. E. Tebeaux, Technical Writing for Women of the English Renaissance: Technology, Literacy, and the Emergence of a Genre, *Written Communication, 10*:2, pp. 164-199, 1993.

9. E. Tebeaux, From Orality to Textuality in Accounting Books: 1556-1680, *Journal of Business and Technical Communication, 7*:3, pp. 322-359, 1993.

10. E. Tebeaux and M. M. Lay, The Emergence of the Feminine Voice, *Journal of Advanced Composition, 15*, pp. 53-82, 1995.

11. W. J. Ong, *Orality and Literacy: The Technologizing of the Word*, Methuen, London, 1982.

12. E. I. Einstein, *The Printing Press as an Agent of Change: Communications and Cultural Transformations in Early-Modern Europe*, Cambridge University Press, Cambridge, England, 1979.

13. S. M. Halloran and A. N. Bradford, Figures of Speech in the Rhetoric of Science and Technology, in *Essays on Classical Rhetoric and Modern Discourse*, R. J. Connors, L. S. Ede, and A. A. Lunsford (eds.), Southern Illinois University Press, Edwardsville and Carbondale, Illinois, pp. 179-192, 1984.